I0050632

I had the pleasure of serving as President of iRobot when Roomba was conceived, developed, and brought to market. I can attest first-hand to the dedication, extreme hard work, and brilliance of the core design team. Borrowing from noted anthropologist Margaret Mead's famous quote, I believe that 'a small group of thoughtful and committed engineers and scientists can revolutionize industries; indeed, it is the only thing that ever has.' This book can serve as a great resource for those who want to bring a true innovation to market, a fun read for those who love Roomba, or just a great story about a hardworking little vacuuming robot.

Helen Greiner, *co-founder of iRobot and serial entrepreneur*

Dancing with Roomba offers an awe-inspiring glimpse into the birth of consumer robotics. A must-read for anyone fascinated by the rise of intelligent machines in our daily lives.

Daniel H. Wilson, New York Times *bestselling author of*
Robopocalypse

I love this book, couldn't wait to get to the next wisecrack or the next disaster for this scrappy team of seven, who pulled off a miracle after so many others had failed. There's wisdom here, along with a good yarn, ending in an SNL skit.

John Mather, *Nobel Prize in Physics winner, 2006*

Joe Jones has written *The Soul of a New Machine* for robotics, providing insight into the long process of bringing the first commercially successful home robot, the iRobot Roomba, to market.

Holly Yanco, *Professor of Computer Science, University of Massachusetts Lowell, Director of the NERVE Center*

Dancing with Roomba is a rare window into the messy, emotional, and deeply human process of innovation. This standout story is a masterclass in creative problem-solving and principled design … Technical readers will appreciate the engineering details; general readers will be drawn in by the wit and storytelling. A must-read for product innovators.

Jeanne Lim, *founder and CEO of beingAI, co-creator of Sophia the Robot as former CEO of Hanson Robotics, committed to advancing humanistic AI for social good*

Dancing with Roomba is a front row seat to the very human story of how Roomba came to revolutionize home cleaning. The book offers a model for how real innovation happens—changing the world requires a small focused team ready to endure a messy grind of overcoming skepticism and constant failure.

Eric Paley, *Secretary for Economic Development of the Commonwealth of Massachusetts*

Dancing with Roomba

Dancing with Roomba tells the unexpected story of the world's favorite robot.

For five decades, corporations spent millions pursuing a floor-cleaning robot. Legions of engineers toiled, dozens of patents were issued, yet every effort failed. Then came Roomba. Selling nearly 50 million units since its launch, Roomba's unlikely success sprang from a breakthrough at an MIT robotics lab, an inventor who persevered through years of setbacks, and an inspired team that worked as one to smash every problem. This book provides a rare view behind the scenes, revealing how a revolutionary product came to be, how it works, and how a tiny company was able to best a crowd of corporate giants.

Written in an easy, narrative style, *Dancing with Roomba* is accessible to all. Anyone who owns a Roomba, works in technology, dreams of building a product, or is just curious about the robot that launched a million memes will find much to love.

Joseph L. Jones, a lifelong tinkerer and inventor, focuses on the practical application of robotic technology to real-world problems. In that pursuit he co-founded Tertill Corporation, a manufacturer of robots that weed home gardens, and Harvest Automation, Inc., a company that builds autonomous robots for the nursery and greenhouse industry. At both firms he served as CTO. Earlier, as senior roboticist, he spearheaded the development of Roomba, the world's first successful home floor-cleaning robot. A graduate of MIT, he is author of three earlier books on robotics and holds over 80 US and several international patents.

Dancing with Roomba
Cracking the Robot Riddle
and Building an Icon

Joseph L. Jones

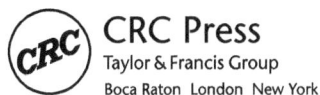

CRC Press
Taylor & Francis Group
Boca Raton London New York

CRC Press is an imprint of the
Taylor & Francis Group, an **informa** business

Designed cover image: Joseph L. Jones

First edition published 2026
by CRC Press
2385 NW Executive Center Drive, Suite 320, Boca Raton FL 33431

and by CRC Press
4 Park Square, Milton Park, Abingdon, Oxon, OX14 4RN

CRC Press is an imprint of Taylor & Francis Group, LLC

ISBN: 978-1-032-88952-8 (hbk)
ISBN: 978-1-032-89061-6 (pbk)
ISBN: 978-1-003-54048-9 (ebk)

DOI: 10.1201/9781003540489

Typeset in Minion
by SPi Technologies India Pvt Ltd (Straive)

To the memory of
Paul D. Fergeson
who handed me the world

_____

Contents

Preface

INVENTING A ROBOT IS like parenting a child. Both are acts of creation that demand commitment and hard work, make you vulnerable to circumstances beyond your control, and deliver immense satisfaction. The objects of parenting and inventing tug at the same emotions—pride, embarrassment, worry, and hope. Both can be all-consuming.

In that spirit, I'd like to tell you the story of Roomba, an electromechanical offspring of mine. A mind-child that, although troublesome and trying at times, ultimately filled the hearts of everyone involved in its birthing with joy and pride. Produced by iRobot, Roomba is an unassuming little robot whose only job is to clean domestic floors. That sounds simple, like the sort of thing that might have been avant-garde in the 1970s. But it wasn't simple, or easy, or quick. Starting in the 1950s, fertile minds from around the world conceived and incubated scores of floor-cleaning Roomba-siblings. All but Roomba perished during infancy. The book's first chapter tells a few of those sad stories.

I'm a life-long tinkerer, an unlaunched physicist, and an obsessed roboticist. I like to invent stuff. The most fun I can have is to dream up some new, hopefully useful thing and then try to bring it to life. Roomba let me exercise that passion more satisfyingly than any other project in my 40-plus years in robotics. It also delivered the most sublime success. Of the many talented individuals who facilitated Roomba's gestation—the robot's mind-parents—I am just one. But it was my honor to play the earliest role. As the story unfolds some of the others, specifically the members of Roomba's core development team, will be introduced in the order they joined iRobot. (Besides the author, the team included project manager Winston Tao, mechanical engineers Paul Sandin and Eliot Mack, electrical engineer Chris Casey, software engineer Phil Mass, and administrator Sara Farragher.)

With a few excursions before and after, the book tells the story of Roomba between 1986 and 2004. In the former year, a groundbreaking new approach to robot-focused artificial intelligence (AI) was developed at the MIT Artificial Intelligence Laboratory (AI Lab) where I then-worked. Based on that idea, I built Roomba's earliest direct ancestor for a Lab-sponsored event in 1989. Over the next ten years, I tried and failed twice to transform Roomba into a product. Along the way I and iRobot (my employer after the AI Lab) learned lessons that would be crucial to Roomba. In 1999 a colleague and I proposed Roomba. As a result, a team formed and we spent three years developing the product. Launch was in 2002, rapid growth followed, and a nationally televised event in 2004 cemented Roomba's place in the zeitgeist.

The book ends with an epilogue and an appendix. The former explains why Roomba survived while its siblings did not. Robots are artificial intelligence embodied, and the appendix delves into how the current astounding advances in AI may affect robotics.

The name "Roomba" was worked out by executives at iRobot and a commercial branding firm. The robot's motion inspired its designation. Namers noted the movements it performed as it went about its task, now skirting a wall, now bouncing away, now pirouetting. To them it appeared that the robot covered the room in a sort of intricate dance. So the namers created a portmanteau of a dance, rumba, and room, forming Roomba. It seemed fitting. I feel like I've been dancing a marathon with that little robot.

I hope you'll enjoy Roomba's story. It kept me on the edge of my seat for over a decade! But more than that, I hope this twisted tale will illuminate how real products, especially revolutionary ones, come to be. The underlying setbacks and triumphs on the road to product-hood are rarely revealed by those involved, so you'll be forgiven for assuming that developing new technology is a dull, straightforward, soulless affair. The opposite is true. Creating a new product like Roomba is a thoroughly human enterprise driven by human emotions. And, like any human enterprise, success requires vision, passion, commitment, and luck. Stir those together and you'll always get a good story. Thanks for giving this one a chance.

And one more thing. There's a companion website for the book that contains many relevant photos, a timeline of Roomba and other robot vacuums, and some additional material. You'll find all that at www.DancingwithRoomba.com.

Joe Jones
Acton, Massachusetts

Prologue

WINSTON, OUR PROJECT MANAGER, wants a quick test of Roomba's final prototype. So, with all seven members of our core development team gathered around, Paul scatters some standard test dirt (broken Cheerios) on the carpet. Chris aims the robot toward the mess and presses the start button.

The expected behavior, practiced by a long series of earlier versions, is for Roomba to advance and gobble up the detritus. The familiar startup sounds play as Roomba drives forward and passes over the dirt. The dirt remains behind, undisturbed.

Sara is first to break the stunned silence. "Is it supposed to do that?" she asks.

It is not! We look at each other, incredulous. We try again. And again. With each attempt our confusion and consternation grow. But Roomba resolutely refuses to clean.

For a moment, Roomba occupies my total focus, I hear and see nothing else—as if intense concentration might spur our failing robot to work. Mind racing, I'm desperately searching for a forehead-slapping moment, one where I can say, "Oh, of course it doesn't work, we forgot X!" But I'm at a loss, there is no X.

Our simple test has laid bare a deal-breaking flaw. And it comes at the worst time. In two weeks the factory begins the birthing process for thousands of Roombas—contracts are signed, deadlines chiseled in stone. Yet today our robot can't do its only job.

Everything hinges on what we do next. Unless we make sense of Roomba's obstinacy and correct it immediately, we are ruined. Thirty months of toil, millions of dollars, and our team's most cherished hopes and dreams will evaporate—just as they have for so many before us. Robotic floor cleaners have been the Bermuda Triangle of technology teams for decades. We thought we'd charted a safe course across that expanse, but suddenly our confidence is profoundly shaken.

CHAPTER 1

Robot Vacuum Dreamin'

THE YEAR WAS 1957, light from the dawning Space Age suffused the horizon, and wonder was on tap. Scheduled for unveiling on an early January day were the latest innovations in home convenience. In living rooms across America families gathered and 15 million black and white TVs flickered to life. Whirlpool Corporation had declared war on drudgery and in a national broadcast they were about to reveal their secret weapon: the "RCA Whirlpool Miracle Kitchen."[1,2,3]

The captivated audience watched as an elegant spokesmodel glided to the heart of the spacious, ultramodern kitchen. There, while extolling Whirlpool's vision, she began demonstrating an arsenal of futuristic labor-saving tools, soon to be commonplace. Among them was an electronic oven that cooked food in the blink of an eye, an autonomous cart that not only transported the family's meal to the table but washed and stored the dishes as well, and a closed-circuit TV that, at the touch of a button, let you check on the baby or see the visitor at your front door.

"Even the floor is cleaned electronically!" the hostess proclaimed. During the grand finale she showed how. Intrigued viewers followed a "self-propelled floor cleaner"—a robot!—as it emerged from its hutch beneath the counter and rushed about the kitchen. Vacuuming up particles of dirt it left the linoleum spotless. This automatic cleaner even boasted a mechanism that let it wash and rinse the floor. Returning to its station afterward, it would discharge the collected dirty water and recharge itself, ready for its next job. Drudgery dispelled, let the bridge game begin!

The TV audience likely viewed Whirlpool's floor-cleaning robot as an inevitable progression of then-burgeoning, postwar consumer technology.

DOI: 10.1201/9781003540489-1

The show stoked the idea that, just like flying cars, the arrival of floor-cleaning robots could be expected any day now. Contemporary works of fiction amplified that message.

ROBOT TALES

In 1950, a few years before Whirlpool's kitchen became a miracle, visionary science-fiction author Ray Bradbury published a short story titled "There Will Come Soft Rains." In it he described a swarm of robotic mice that emerged each day to scour clean the floors of a smart house. Poignantly, this activity continued long after the family whose home it had been perished in war.

A pioneering television series called *Science Fiction Theatre* dramatically spotlighted a floor-cleaning robot in 1955. In an episode titled "Time Is Just a Place" suburban homeowner Al Brown became curious about the couple who had recently moved in next door. Peeking through their window after ringing the doorbell, Al spied a small hemispherical device with glowing indicator lights and a rotating antenna zipping about the floor. Whenever it passed over a bit of dirt, the dirt vanished!

When neighbor Ted Heller finally answered the door, he explained to Al that he had invented a robotic floor cleaner called the "Sonic Broom." Al rushed home to tell his wife Nell, "Honey, you can throw away your vacuum cleaner!" Soon, Al was certain, robot cleaners would spread across the nation to ease the burden of every American housewife.

In 1956, the year before Whirlpool's presentation, iconic author Robert Heinlein made a little floor-cleaning robot a central feature in his novel *The Door into Summer*. Heinlein's protagonist explained why his robot achieved wild success. Even though the robot was not a very sophisticated machine, it sold well because the robot's price matched that of manual vacuum cleaners. (Later we'll see how remarkably prescient Heinlein's reasoning was!)

These touchstones set the stage. By the 1950s the robot vacuum cleaner was a familiar idea, its popularity was assumed, and contemporary technology looked ripe for its creation. Surely, all inventors needed to do was to pick the right components, apply a little ingenuity, and then rake in eye-popping profits from an eager market.

They tried.

MIRACLE ROBOT

It turns out that Whirlpool's Miracle Kitchen robot was both science and fiction. For purposes of directorial expediency, the device filmed for the TV spot was remotely controlled by a stagehand positioned off camera[4]—it wasn't really a robot.[5] But authenticity would have been possible—Whirlpool had the means to demonstrate a truly autonomous machine. In one of the earliest, serious efforts to build a robot floor cleaner, Whirlpool engineer Donald G. Moore had recently invented a credible machine he called a "perambulating kitchen appliance."

Moore's patent[6] described a diminutive, square-shaped, three-wheeled robot maybe six inches tall. The robot included a mechanism for cleaning the floor, an electric motor for propulsion, and one tricycle-like wheel for steering. Moore devised a strategy that let the robot cover—and thus clean—the entire floor. So, what steered the guide wheel to implement that strategy?

Today we'd use a computer, but here Moore was out of luck. The closest candidate in the year of the Miracle Kitchen was likely the IBM 610, a vacuum tube-filled beast the size of a desk that weighed 800 pounds and drew over two kilowatts of power.

Rather than a computer, Moore relied on a simple mechanical strategy to steer his perambulating appliance. The robot's single guide wheel was mounted on a pivot that let it steer left or right. Connected to the pivot was a lever arm. At the end of the lever was a magnet in close proximity to the floor. Inlaid into the floor was a long, thin ribbon of ferrous metal closed in a circuitous loop. The robot started with its steering magnet hovering above the metal guide. Then, as the robot moved forward, magnetic forces kept the magnet—and thus the steered wheel—aligned with the metal ribbon. The layout of the ribbon therefore defined the robot's path through the kitchen.

The steering scheme was quite clever, but it shackled the robot to a fixed, unchangeable route. Pushing a chair back from the table after a meal, for example, risked confounding the robot—the out-of-position chair might block the robot's inlaid path. Moore's patent disclosed an instrumented bumper that halted the robot when it touched an errant chair leg or other obstacle. But that was its sole strategy. Once halted, human intervention was needed to reposition the chair and restore the regimented order this 1950s robot required.

This and several other issues rendered Moore's robot impractical as a consumer product. Whirlpool Corporation never attempted to commercialize it. Relying on an embedded magnetic strip to tell the robot where to go was deemed unworkable even then.

TRY, TRY AGAIN

In the decades that followed, the dream of the robot floor cleaner proved enduringly seductive. Many hats—worn by inventors, entrepreneurs, and corporate captains—were tossed into rings. The first machines, like Donald Moore's perambulating appliance, utilized mechanical intelligence, later ones were directed by computers. But regardless of their means of control, each attempted robot became a melancholy tale of inspiration, hope, and hard work culminating in disappointment. No viable product reached its intended destination: customers' dirty floors.[7]

SIEMENS

Dr. Fritz Kaiser, PhD, was a seasoned employ of Siemens, the German multinational corporation, when he took up the robot vacuum challenge. Having previously tackled fluorescent lighting control, water-tank testing, and microwave ovens, Dr. Kaiser brought an eclectic sense to his new task. The patent he was granted in 1962 describes an ambitious and innovative design. Kaiser's robot featured a clever arrangement of mechanical switches that activated an electric steering motor. Guided by the switches, the robot could follow walls and turn away from obstacles it encountered. Neither computer nor inlaid path was required.

But the robot's most remarkable feature was its strategy for efficient cleaning. Like a lawnmower tracking the edge of the cut grass, Kaiser's robot tried to follow the border between the dirty parts of the floor and the areas it had already cleaned. It achieved this by splitting the vacuum nozzle into two parts and using light sensors to monitor the relative dirtiness of the air flowing through each half—it steered to keep the dirt levels at a preset ratio.

Sadly, no commercial rewards came to Dr. Kaiser's daring machine. Methods that are appealing on paper and even show promise in the lab often come to grief in the real world. Today the patent office[8] is the Kaiser robot's only memorialist.

BUMP'N GO VAC

Queens, New York resident Paul H. Ku was a veteran inventor. He had already patented a solar umbrella able to warm a cup of coffee and a portable air conditioner that kept its user cool when he turned his attentions to the robot vacuum cleaner problem. The key to success, Ku believed, was a contrivance used by countless mobile toys—a *bump-and-go* mechanism. Such a device consists of an inspired arrangement of gears and wheels that causes the mechanism to change direction anytime it encounters significant resistance along its current direction of travel.

Ku's 1979 patent[9] describes an "Automatic vacuum cleaner" with three supports. Two were passive ball casters capable of rolling in any direction, and the third was a bump-and-go mechanism that provided direction and motive force. Just like the toy, but on a larger scale, the bump-and-go mechanism would cause Ku's vacuum to change direction when it bumped into obstacles. Theoretically this let it bounce around a room until the floor was clean. The robot included a battery that powered both the drive mechanism and a conventional brush/vacuum combination to pick up dirt. No computer was needed.

But Ku's robot never made its commercial debut. The patent waved a couple of red flags. First, as we'll see later, running a conventional vacuum mechanism from a battery forces the designer to choose between a huge battery and a minuscule runtime. Second, bump-and-go mechanisms can easily be stymied by uneven surfaces or transitions between hard floors and carpets. These issues likely made Ku's robot impractical.

ROBBY

Hitachi Corporation, a leading Japanese conglomerate, embarked on their effort to revolutionize home cleaning in 1983. Executives directed a team of top engineers to build a robot vacuum. By 1985 several patents related to a "Self-traveling robot" known as "Robby" had been filed.

The pride of its creators, Robby was a substantial machine, about 31 inches long and 20 inches wide and tall; it used a standard vacuum to pick up dirt. Resplendent with the latest technology, the robot included a sonar distance sensor, a gyroscope, and other sensors, all answering to an onboard microprocessor. At the zenith of Robby's renown a prototype was showcased at Domotechnica, an international trade fare held in Germany in 1985. Afterward, although a few more patents were granted, Robby wandered into obscurity.

While cleaning, the robot made a valiant attempt to visit every point in a room by following a tidy pattern known as a *boustrophedon*[10] path. But the external features the strategy relied on may have been Robby's undoing. One patent[11] tells us, "If there is an obstacle in the room, the area after it becomes an uncleaned area, and if the room is not rectangular, an uncleaned area may exist." That is, for Robby to work well in their homes, customers would have to forgo the luxury of furniture.

SANYO

Major corporation Sanyo, bitten by the same bug as Hitachi, decided that they should be the one to bring the convenience of robotic floor cleaning to the masses. So, in 1986, the year following Robby's public introduction, Sanyo filed a first patent[12] for a "Self-propelling cleaning robot." Additional patents followed.

Setting themselves apart from other contenders, Sanyo's robot sourced most of its power not from an onboard battery, but through an electrical cord that unspooled from the robot and connected to an electrical outlet on the wall. This solved a tricky power problem but created an even trickier cord-tangling problem, which likely proved intractable in real-world environments. In any case, like its predecessors, the Sanyo floor-cleaning robot lived out its lonely life in a laboratory.

TELEVAC

In 1987 the Jonas brothers—French inventors Andre and Barnard—patented[13] a unique and imaginative design that might be considered the chameleon of robot vacuums. The description was apropos, not because their Televac robot could change colors, but because it could shoot its "tongue" out across the floor to grab dirt.

Televac was a cylinder maybe two feet tall and a foot and a half in diameter. It cleaned rooms from the sidelines—always hugging the nearest wall as it moved. While gliding along Televac would make frequent stops. When it stopped a vacuum head attached to an 18-foot-long hose wound into a reel would play out from beneath the robot. The head sucked up dirt as it snaked into the room. At the end of its travel, or when the head bumped into an obstacle, the hose would reverse and retract completely. Then the robot would rotate a few degrees, and the head would reach out again. Following this strategy, the room would eventually be cleaned—barring certain unfortunate arrangement of obstructing furniture.

The brothers founded a company called Azurtec to commercialize their invention and garnered some significant publicity at the time.[14] But Azurtec sold no machines, and another robot vacuum quietly disappeared.

PANASONIC

Panasonic poured its corporate heart into crafting a practical robot vacuum. The resolute, dozen-year quest began in 1990 when five engineers planted the company flag in robot vacuum land by patenting a "Self-running cleaning apparatus."[15] The long-running development was chronicled by occasional news stories. At one point, test marketing was reported to have commenced in Japan with the robot's price expected to be somewhere north of $2000. Intriguing photos were published of an attractive, squarish, transparent blue robot with rounded corners. It stood about a foot tall, bristling with over 50 sensors of various kinds. In the last apparently related patent published in 2002[16] credit was shared among 13 inventors. But despite the sustained, well-staffed, and no doubt costly effort, Panasonic's robot only broke the hearts that held it dear.

HomeR

Frank Jenkins, an American inventor and computer programmer, was a member of a small group of robotics enthusiasts in northern California at the time he built his "HomeR Hoover" robot.[17] Frank's colleagues were impressed when he demonstrated his attractive robot at a meeting of the group. So was *Discover* magazine,[18] reporting on the robot in 1993.

Standing two feet tall, 1.3 feet wide, and weighing 44 pounds, HomeR was a truncated pyramid hosting 80-plus sensors of various sorts. Jenkins liberated components from two different Black & Decker vacuums to construct HomeR's cleaning mechanism.

When vacuuming finished, HomeR's sensors and software enabled the robot to return to its charging station and refresh its batteries. Comparative performance details are lacking, but from appearances, built-on-a-shoestring HomeR seems no less capable than the robot vacuums developed by corporate giants of the era. But, mirroring his larger counterparts, Jenkins' HomeR never found the entrance to the marketplace.

TOMY DUSTBOT

One robot that *did* reach market was Dustbot. In 1985, Japan's Tomy toy company released an autonomous robot designed to vacuum detritus from the smooth, flat surface of a table or desk. Powered by a single motor in a four-inch-tall enclosure, the robot clutched a miniature broom that, adorably, whisked back and forth as it moved. Navigation was accomplished using an ingenious combination of passive rollers, driven wheels, and gears. This let Dustbot turn away from the edge of the table, and also let it escape from any obstacle it bumped into.

The robot's motion was fully determined by mechanical interactions. Because it lacked a randomizing element, Dustbot sometimes settled into an unproductive, repeating pattern. Although intended as a novelty, Dustbot is significant because of the key lesson it taught—it revealed what a robot vacuum *didn't* need. At a very low price, Dustbot accomplished its mission without cameras, scanning lasers, sonar rangers, precise maps, probability theory, motion planning algorithms, or any of the other advanced tools technologists and academics of its day typically stipulated as essential. Long before Roomba, Dustbot showed by example that a simple, low-cost process can solve the *coverage problem*—ensuring the robot visits every part of the floor. That Dustbot was limited to smooth desktops was incidental.

ASIMOV'S MIRAGE

For science-fiction authors, humanoid robots offer immense potential as plot devices and many have explored their dramatic possibilities. None were more prolific or influential than Isaac Asimov. In Asimov's universe, robots are bound by three laws[19] designed—like safety systems embedded in industrial machinery—to keep people safe from harm in the presence of robots. Never mind that implementing the laws assumes an extraordinary level of cognitive agility that even today's robots seem far from mastering. The conceit of the three laws has generated a wealth of great stories. Asimov's 1950 anthology *I, Robot* contains several; its title inspired the company name, iRobot. (In that case the "i" stood for internet.)

But Asimov's beguiling vision conceals a subtle snare—the assumption that robots should be "general-purpose." That is, unlike crass appliances, a proper robot should not be designed to solve a particular problem but rather should be endowed with the ability to solve *any* problem, to accomplish any task that a person might do. After its manufacture, training or programming could specialize the robot to a particular task. The author's stories predicted that industry would first develop general-purpose machines; only later would (boring) robots dedicated exclusively to specific jobs emerge.

And thus were generations of aspiring roboticists lured toward marketplace failure.

In general, building a robot mechanically and sensorially capable of matching all human abilities is fantastically more difficult, more expensive, and less likely to succeed than building a device to perform a specific task. Consider the approach to automatic cleaning—once frequently suggested—of first perfecting a humanoid robot and then having it push a manual vacuum cleaner about the house. Pursuing the ideal of general-purposes-ness makes solving difficult problems many times harder.

And yet, one of the most widely replicated mobile robot company business plans has been to build—as nearly as possible—a general-purpose robot. This done, the immensely challenging task of making the robot do something useful was left as an exercise to the customers.

The one case where such robots, also known as *platform robots*, have seen modest success is when the targeted customers are educators, researchers, or hobbyists. For this contingent, the robot *is* the task. But the track record of general-purpose robots applied to the job of floor cleaning has been a poor one.

RB5X

Revered RB Robotics CEO Joe Bosworth brought to life one of the first commercially available platform robots in 1982.[20] His RB5X robot, reminiscent of *Star Wars'* iconic R2D2, was priced at about $1500—relatively modest for the time. It quickly aroused the passions of hobbyists and educators.

The RB5X included sensors for measuring range, touch, and light. It could be programmed via a connection to the user's computer. And, true to form for a general-purpose robot, out of the box it could carry out no useful tasks.

But in 1983 the company developed a vacuum accessory that attached to the robot and enabled it to clean floors. Unfortunately, the vacuum increased demand on the battery, significantly reducing robot run time, and it changed the geometry of the robot in a way that created navigation problems. The new module never went into production, and another assault on robotic floor cleaning was routed.

CYE

Serial entrepreneur Henry Thorne launched Probotics Inc. in 1997. He'd invented[21] the company's main product, a platform robot called Cye and was eager to get it into customers' hands. Priced at about $700, Cye was popular among hobbyists. A squat 16 inches wide by 10 inches deep, the robot was about the size of a toaster oven. It sported two spiky, rubber-coated wheels, one on either side of the robot. The spikes enhanced the robot's navigation proficiency on carpets. A small caster wheel centered behind the drive wheels gave the robot balance. Included standard with Cye was an "upright vac attachment" that enabled the robot to tow a cordless vacuum behind it. But despite benefiting from a clever navigation stratagem,[22] Cye's performance was not generally good enough for real world operations, and customers with floors to clean did not flock to Cye.

CAREBOT

Texas-based Gecko Systems was founded in 1998 by Martin Spencer. With the admirable notion of providing care for the elderly, Gecko marketed their vaguely humanoid, four-foot tall CareBot robot.[23] Like many "personal robots" of the era, CareBot's datasheet included a long list of technical features. But the list of useful tasks it could perform out of the box was empty. CareBot was, however, available in a vacuum-ready configuration. For an additional $300 over the base price of $2,295 the robot would be supplied with an integrated vacuum cleaning system—but no programming. That puzzle was left to the buyer to solve. CareBot cleaned few floors.

STATE OF THE ART

In the half-century before Roomba, many corporations, university researchers, and basement-based tinkerers lavished thought, resources, and time on robot floor cleaners. More than a score of would-be robot revolutionaries received patents[24] related to robot floor cleaners they had developed.

Why were so many companies and individuals so fixated on building a robot vacuum? Mostly for two reasons: market need and technical feasibility. First, we all have floors and almost none of us enjoy cleaning them. So, a machine that could relieve consumers of that chore seemed likely to be highly popular. Second, the task looks alluringly simple. Of all the jobs we might delegate to a robot, floor cleaning is perhaps the least technically challenging. A robot floor cleaner doesn't need dexterous hands as might be required were it tasked with putting away dishes or folding clothes. It has no use for complex, expensive arms as it would were it required to lift heavy things or reach high up to clean a window. It doesn't even need clever vision sensors to recognize people or determine whether a spot is dirty or clean. All the robot need do is tote a cleaning mechanism around the floor for long enough to be sure the floor is clean. That is, it just needs to move around without getting stuck—how hard could that be?

Deceptively, frustratingly, infuriatingly hard is the answer. Beyond technology, a viable product must successfully address issues of cost, size, performance, and convenience among others. And so for decades, every one of the many attempts to build and market a robot vacuum cleaner—no matter how inspired or well-funded—ultimately proved disappointing. Each robot so lovingly conceived by its mind-parents, programmed with such hope, and tested amid great expectations, in the end broke its creators' hearts. Most robots never left the lab of their birth, the few that did failed to thrive and stand on their own.

These non-successes are all the more poignant because inventors of the real robot vacuums described above—along with many others—could all claim their machines "worked." Each could set up circumstances where their robot moved about, avoided obstacles, and picked up dirt. That is, every inventor could perform a convincing demonstration.

But a demonstration is not a product. And every robot vacuum before Roomba contained at least one fatal flaw. Some were too big or too heavy to be practical. Many used navigation schemes that couldn't reliably cover the area they were supposed to clean. Others were immature, requiring the user to write a program or provide the robot with a map of its surroundings. If a price was specified, it was always far too high—never aligned with the price of manual vacuum cleaners.

These conclusions are so evident in retrospect! Today, with the market for robot vacuums now well established, customer surveys, buying patterns, and so on make clear which features consumers value. But in the 1990s and early 2000s no one knew. And asking potential customers couldn't help—all answers are speculative when the product is theoretical. To a distressing degree, all of us would-be robot revolutionaries were guessing.

The biggest corporate contenders, the likes of Siemens, Hitachi, and Panasonic, appreciated the shortcomings of their prototypes even then. And despite the millions they may have invested in development, all declined to bring to market a product that wasn't ready. But by the mid-1980s the technology *was* ready. The delay in the emergence of robot vacuums for another decade and a half can perhaps best be attributed to matters of imagination.

NOTES

1 *Broadcasting Telecasting* magazine, 1/7/1957, p. 93: https://www.worldradio history.com/Archive-BC/BC-1957/1957-01-07-BC.pdf.

2 Video: https://www.youtube.com/watch?v=Vui2CSEwOxQ.

3 The set of the Miracle Kitchen would go on to be reproduced at the American National Exhibition held at the height of the Cold War in Moscow, USSR in 1959. There it would form the backdrop for the "Kitchen Debate," an impromptu sparring session between US Vice President Richard Nixon, touting the achievements of US capitalism, versus Soviet Premier Nikita Khrushchev, who extolled the successes of his country's communist system.

4 Indy Week, 6/5/2013, https://indyweek.com/food-and-drink/secrets-behind-1950s-miracle-kitchen-future/.

5 If the filmed device wasn't a robot, you may wonder, then what exactly is a robot? The question bedevils all who ponder it. Is a teleoperated arm a robot? How about a dishwasher? Ambiguity arises from the fact that "robot" is not a well-defined term constructed by technologists. Rather, it is a literary locution coined by playwright Karel Čapek for his 1921 play *R.U.R.* (Rossum's Universal Robots). On literary matters, universal agreement is rare—all opinions being

welcome. My opinion is that, to qualify as a robot, a machine must operate in a world where surprises can occur and must respond to them competently on its own. Devices not passing this test (like teleoperated arms and dishwashers) are servomechanisms. But other opinions abound.

6 Patent US3010129A was filed in in 1957 and granted in 1961.

7 See, for example, E. Prassler, et al., "A Short History of Cleaning Robots," *Autonomous Robots* 9, 211–226, 2000.

8 German patent DE1239819B was applied for in 1962.

9 US patent US4173809A.

10 A boustrophedon path is a common and efficient means for covering an area. This raster pattern connects each row to the next by alternating ±180 degree turns (either square or rounded). The term, a combination of Greek words meaning "as the ox turns in plowing," has been in use since the 1600s.

11 Japanese patent JPS62154008A. Subsequent work may have made the robot more furniture friendly.

12 Japanese patent JPS62170219A.

13 European patent EP0274310B1.

14 *Popular Science*, October 1992.

15 US patent US5109566A.

16 European patent EP1265119A2.

17 F. Jenkins, "Practical Requirements for a Domestic Vacuum-Cleaning Robot," AAAI Technical Report FS-93-03, 1993.

18 *Discover* magazine (March 1993: 28).

19 Asimov's three laws: (1) A robot may not injure a human being or, through inaction, allow a human being to come to harm. (2) A robot must obey orders given it by human beings except where such orders would conflict with the First Law. (3) A robot must protect its own existence as long as such protection does not conflict with the First or Second Law.

20 Asimov, I., & Frenkel, K., (1985). *Robots, Machines in Man's Image*. Harmony Books.

21 Patent US6046565A.

22 Cye attempted to maintain positioning accuracy by automatically driving to a known calibration point and recalibrating anytime the robot determined that uncertainty had grown too large.

23 https://web.archive.org/web/19990424033940/http://www.geckosystems.com/carebot/rst.htm.

24 Beyond those already mentioned, selected companies and individuals, followed by year granted and patent number, include: Knepper Hans Reinhard, 1985, US4700427A; Mitsubishi Electric Corp, 1989, JPH02249522A; Gold Star Co, 1990, US5307273A; NEC Home, 1991, JPH0546246A; Guy T. D. Ashworth, 1991, US5321614A; Jin S. Hwang, 1992, US5568589A; Samsung Electron Co, 1992, US5369347A; Andre Colens, 1994, US5787545A; Nippon Yusoki Co, 1994, JPH0889451A; Ajit P. Paranjpe, 1995, US5634237A; Kwangju Electronics Co, 1995, US5839156A; Minolta Co, 1996, US5894621A; Honda Motor Co Ltd, 1997, JPH11178765A; Timothy P. Allen, 1997, US5995884A; Erwin Prassler, 1998, DE19849978C2; Volker Sommer, 1998, US20010004719A1; Dyson Limited, 1999, WO2001006904A1.

CHAPTER 2

Passion, Desire, and Robot Programming

I WAS BORN A NERD.

The earliest evidence comes from a story my mother liked to tell. It seems I embarrassed her on every visit to the pediatrician. While waiting for the doctor, I liked to crawl around the edge of the floor. Whenever I spotted an electrical outlet in the wall I'd point and delightedly shout, "Pug-in! Pug-in!"

Other childhood events further confirmed my nature. One morning I saw some people on TV. They were holding a model of a satellite the United States had just launched into orbit. Everyone was excited and happy. It looked like so much fun that I decided to start a space program of my own.[1]

To construct my spacecraft, I rolled a newspaper into a cone and secured it with tape. Inside the cone I inserted two D-cell batteries to supply electricity and a squat, empty can of Lux dishwashing liquid to hold the rocket fuel. I mixed the propellent myself (it was mostly water).

The mental picture I'd formed of the launch could not have been clearer. My satellite would blast off from the top step of our back porch, zoom above the trees lining the dirt road behind our house, and then continue climbing, quickly receding to a small bright dot gliding across the night sky.

On a winter evening not long after sunset I completed construction. I carried my satellite outside, positioned it on the back steps, and had my mother insert a lit match into the fuel tank. (Some subtleties of rocket propulsion escaped me at the time.) Because of the cold weather we then retreated inside. Later, I expectantly peeked out the door see how the launch had gone.

My satellite was just sitting there.

"Devastated" doesn't begin to describe how I felt. The failure of my newspaper spacecraft to achieve earth orbit was by far the most crushing disappointment I had ever experienced. Heartbreak overtook me and for a few minutes I became inconsolable. I was five years old.

PARSIMONY TRAINING

Temporary setbacks notwithstanding, my fascination with all things science and technology only deepened. As I grew, I pestered my mother to bring home books from the library containing experiments that I could do. She often humored me. I petitioned my parents for a chemistry set. They made me wait until age 10. I disassembled mechanical clocks I'd been given and, later on, our old television. I used the controller from my model train set to supply power for electrical experiments. I tried to build a transistor and a computer. Both were a bust. But the buzzer made from a coil I wound myself and an electrode snipped from a tin can did work! Whenever possible I acquired magazines, mostly about electronics, that contained projects I could do. At age 14, one taught me how to build a one-vacuum tube AM radio transmitter. Thrillingly, that also worked!

My experiments were not without risk. Once I hooked up the *flyback transformer*.[2] I'd removed from our old TV to my train controller. It made a very cool quarter-inch long spark. But then I accidentally touched the output lines. Suddenly I felt like an angry bodybuilder had grabbed me by the shoulders and shaken me violently three times. On another occasion I spent an afternoon investigating the properties of sulfur. The element makes the most beautiful blue flame when burned! Unfortunately, the sulfur dioxide gas produced in the process is not benign. All that evening I felt like an elephant was sitting on my chest. My mother wasn't much interested in science, so I decided not to bore her with these details of my investigations.

Circumstance hampered my research program in a couple of ways. First, living frugally in the tiny rural community of Morrisville, Missouri (population 256), my family and I resided above the poverty line but well below the comfortable mark. Our situation meant that I could rarely go out and buy the things I needed. To enable my experiments and projects, scrounging and making do were the orders of the day. My father was an auto mechanic and some of my supplies I scavenged from broken components he'd removed from cars. Another great source of material was the spot behind a shed where the high school science teacher occasionally dumped broken and unwanted items from his classroom.

A second hindrance was that no one I knew shared my interests. Neither family, nor friends, nor even teachers at school expressed wonder about things that fascinated me—like how distant broadcasts could be received because short radio wavelengths bounce off earth's ionosphere, or how chemical elements could be identified by the spectral lines they produced, or how a voltage could be generated by swinging a magnet past a coil of wire. I learned not to ask questions about such things even of teachers. It was understandable when my elders didn't know the answers, but I found it most disheartening that they seemed not to care about the questions!

My inventory almost never contained all the components needed for an experiment or published project I wanted to do. Being unable to follow the cookbook meant I had to reinvent the recipe. But making substitutions or finding a different approach altogether was possible only if I truly understood the underlying principles. My compulsion to build and experiment led to a lot of self-education.[3]

Never being able to just buy the parts and tools I needed felt frustrating at the time. Scarcity made every project more complicated and more time consuming. But material austerity sowed seeds of innovation. If I couldn't do something one way, there were always other ways, and I learned to find them. That would turn out to be a pivotal skill.

Over the years, as I accumulated knowledge, I passed through several stages of "What do you want to be when you grow up?" Starting with TV repairman (the most exotic job my 10-year-old mind could conceive), I advanced to electrical engineer (the person who designs the TV in the first place), and finally to physicist (the individual who discovers the laws of nature the electrical engineer uses to design the TV). My love of science and technology was steadfast, and I wanted to take it all the way. But I didn't entirely know how.

There were 12 teachers in my tiny, country high school. One of them, the new psychology teacher whose courses I had never taken, waylaid me in the hallway between classes one day. He asked a few questions and then inquired about my plans for college. As valedictorian I'd get a scholarship to the University of Missouri … and that was as far as I'd thought about it. Mr. Fergeson said, "You really ought to consider some other schools like Cornell, or Cal Tech, or MIT."

It was for that reason only that I applied to MIT for college. Arriving there as a freshman in the fall of 1971, amazingly, I found home! At MIT everyone shared my interests. Everyone cared about the questions. And everyone harbored a burning curiosity to learn more.

I maintained my plan for a career in physics through commencement. But once in graduate school, I hit a pothole. Physics, it turned out, wasn't specific enough. I had to pick a narrow subfield for deep study. When Rick Shafer, my best buddy from undergraduate days, pursued his PhD in astrophysics, he focused with such intensity that he often neglected sleep and ultimately lost synchrony with the day/night cycle. An all-consuming fascination with his research topic compelled Rick to these extremes. But I could muster no such devotion to any specific niche in physics. Passion was a degree requirement and mine was missing.

Hoping to discover the requisite fervor, I left grad school[4] and got a support-staff job at a physics research facility.[5] Three years later passion hadn't shown up. Maybe I needed to search farther afield. So, I quit my job and then, traveling solo, proceeded west from my home near Boston. The next 12 months I spent circumnavigating the globe—visiting exotic places, meeting fascinating people, and experiencing unique cultures. But not in the temple in Muktinath, the caves of Guilin, nor Beethoven House in Wellington—or a hundred other venues—did I find my passion waiting patiently for me.

Now it's 1982, and I was back where I started, but with my savings gone. I was borrowing money from friends and, with increasing desperation, looking for a job. On my way home one day I walked past a gas station with a help wanted sign in the window—I thought seriously about applying. But soon afterward, in a copy of *Tech Talk*, the official newspaper of my alma mater, I ran across a posting for a job at the MIT Artificial Intelligence Laboratory.[6] They were looking for staff to support faculty and graduate students doing computer science research. I applied, and in another stroke of life-changing luck, they hired me. Within weeks I discovered that's where my passion had been hiding all along!

IDEAS

The MIT AI Lab was situated in 545 Technology Square, Cambridge. In the tortured logic of MIT's numbering scheme, our building was known as NE43. Each day I walked up the eight steps of the plinth supporting the building and then took the elevator to my office on the seventh floor.

Research labs and the main computer system occupied the ninth floor of NE43. The seventh and eight floors held offices arrayed along a corridor surrounding the central core. At one end of each floor was a large open "playroom" strewn with couches and chairs and dominated by a floor-to-ceiling whiteboard.[7] A beehive of activity hummed around the clock—faculty ruled the day, graduate students the night.

Ideas were the currency of the Lab. Their abundance made the Lab a place of great wealth. Forging and exchanging ideas was ceaseless. In elevators, offices, and playrooms and through seminars, papers, and meetings, ideas were proposed and argued over. Experimentation and computation determined the fate of each idea—whether it was destined for refinement, repair, or abandonment.

Despite the fact that I was but a research staffer—employed to support others—I was encouraged to contribute. And my ideas, I found, were taken as seriously as anyone else's. At the Lab, the strength of the idea mattered; the prominence and status of its proponent did not. That attitude, embedded in Lab culture, was enforced by our director, Patrick Winston. Pervasive collegiality enhanced the Lab's success and the contentment of its members.

While working at the AI Lab it was impossible to be unaware of the latest cutting-edge research. On any given day I could have a casual chat with *the* leading authority in some branch of AI or robotics. I could hang out with top-tier graduate students preparing to lead the next generation. In the playroom I could join animated conversations before the giant whiteboard: What enables consciousness? How can we program a computer to think like a person? How can we make a robot work in the real world?

Years earlier in my college dorm I'd had similar-sounding, late-night discussions. But at the AI Lab it wasn't just talk. The citizens of this realm possessed the vision, technical brilliance, and discipline not just to wonder about the world but to understand and change it. Nothing I'd experienced before came anywhere near the excitement and vitality of life at the AI Lab.

BOLTED TO THE FLOOR

In the mid-1980s the world of robotics was ablaze with promise—mostly unfulfilled. Since the 1960s robots[8] had worked in factories—where their roles were rigid and limited. Robots' lack of perception and common sense meant that only experts could put them to use. Also at that time so-called "personal robots" were experiencing their quarter-hour of fame. Machines with names like Topo, Gemini, and Hero were marketed to consumers.

But beyond entertainment and education, such robots lacked any practical purpose, and they failed to gain traction in the marketplace. As a now card-carrying roboticist, I found that perplexing and disappointing. Similarly, everyone at the Lab was eager to expand robots' abilities.

I was drafted into a four-person research team tasked with developing a novel software system for a particular class of robots. Ask our boss, Tomás Lozano-Pérez, what we were doing, and he'd say, "We're trying to revolutionize American manufacturing." Heady stuff! Our project, "Handey" (for "Hand-eye"), aimed to increase the presence of robots in industry by making robots much easier to use. The work required me to become proficient in new concepts with cool names like A-matrix, Jacobian, and configuration space. I loved it! When Tomás decided we should write a book about Handey, I learned a surprising and useful lesson—that authorship didn't demand superhuman ability; mere mortals could write books, too. Ours was called *Handey: A Robot Task Planner*.[9]

We worked on a type of robot called a *manipulator*, also known as a *robot arm* or *industrial robot*. It was the type of robot a factory might use. Like a human arm, a manipulator had several joints. Usually there was a "waist" joint that lets the entire arm twist left or right, along with shoulder, elbow, and wrist joints that did what their names suggested. At the end of the arm was a gripper (or *end-effector* in robot parlance) that let the robot pick up and manipulate objects.

Although they were clever and useful, the sort of robots my group focused on generated the one glitch in my otherwise idyllic situation. I found manipulator robots unsatisfying. It had to do with the way they operated.

How would you make a robot assemble something? The standard way was this: Feed the needed components into a *work cell*, the bounded space where the robot operated. Use fixtures to make sure every incoming part was held exactly where the robot expected it to be. Then teach the robot an excruciatingly detailed plan for how it should move; e.g., set joint one to 27.6 degrees, simultaneously move joint two to 13.2 degrees, open the gripper, and so on and on and on. This *motion plan* was often tediously taught to the robot by a worker using a specialized sort of remote control called a *teach pendant*.

In those days, a manipulator robot usually had zero knowledge of the world in which it lived. It knew the position of each of its own joints but nothing else. All such a robot could do was to move from one previously taught *pose*[10] to another. Picture a welding robot programmed to join the parts of a car. Now imagine that a small glitch occurs such that the parts come up a bit short of the position the robot expects. The oblivious robot will happily weld the air!

For decades, this was how industrial robots operated. Manufacturers bolted their robots to the floor, confining them to their work cells—isolated, rigidly ordered, artificial worlds where surprises were forbidden. Because human workers are full of surprises, they had to be kept at bay by a cage surrounding the robot. The Handey project added a sensor to this scenario and eliminated the tedium of creating the robot motion plan manually. We figured out how the robot could plan all the monotonous details on its own. In computer science that step is called *motion planning*.

But as proud as I was of our work on Handey, improving manipulator robots felt incremental.[11] Outside the factories was a whole big world filled with things that needed doing. Robots could help in a thousand new ways if only they could throw off the cognitive chains that shackled them in place. My new-found passion demanded revolution, and I craved exactly that. And revolution, I was about to discover, was already in progress, just down the hall.

PLANS? WE DON'T NEED NO STINKING PLANS!

At the MIT AI Lab, not far from the Handey room on the ninth floor of NE43, lay the domain of Professor Rodney "Rod" Brooks. Rod's team of grad students and staffers were known as the Mobot[12] Group. Together they were pioneering an approach to mobile robots radically different from the conventions of the day. This was the revolution I was looking for.[13]

My group's project, Handey, graduated from Ye Olde School of Motion Planning. Our methods were rigorous and meticulous, our plans mathematically guaranteed.[14] The robot control paradigm we subscribed to, called *modeling-planning*, was the then-standard one, applying to both manipulator and mobile robots. Rod's Mobot Group would have none of this. They saw planning not as the solution but as the problem holding back robotics. Proselytizing his new ideas, Rod wrote a paper with the audacious title "Planning Is Just a Way of Avoiding Figuring Out What to Do Next."[15]

I was very much attracted to Rod's new approach—and not just because of the Mobot Group's rebellious irreverence. Rod's robot control paradigm, now called *behavior-based programming*, promised robots capable of working in our world where surprises abounded. That would finally let robots escape their cages (cognitive and physical!) and take on a thousand tasks people wanted done, but didn't want to do themselves.

The difference between my Handey group and Rod's was the difference between planning and behaving. Here's what that means.

PLANNING

Handey, like all robot planners, *modeled* the world. That is, Handey created a virtual world—expressed in mathematical terms—that included the shape, position, and orientation of all the objects within the robot's reach. The information used to build this model could be supplied by the programmer, could be inferred from sensor-based measurement the robot made itself, or both.

Once the model of the world was in place, the programmer would ask the planner to change the world in some particular way. This usually meant one of the objects in the world should be moved from its current position and orientation to a different pose. Assembling a product required many such moves. Because Handey had a mathematical understanding of both the robot and the world it could, using some advanced techniques, figure out exactly how the robot needed to move in order to put the object in its new pose while avoiding collisions with other objects along the way. The plan Handey created could be quite complex, even involving putting the object down and picking it up with a different grasp when necessary.

When it finished planning, Handey directed the robot to begin moving. Importantly, Handey assumed a world frozen in place. From the moment Handey built its model of the world until the robot finished moving, no unknown objects could enter the world, and no existing objects were allowed to budge on their own.

Planning in this way is valuable because it offers guarantees. Ask a planner to move an object from one place to another and only one of two things will happen: (1) If repositioning the object is physically possible, the planner will deliver a complete plan to make it happen. (2) If the motion is physically impossible, the planner will tell you so. Like Yoda, planners proclaim, "Do or do not. There is no try."

The problem with planners (especially in the days when sensors were rare and computers less powerful) was that their plans were brittle. If their world model contained errors, was incomplete, or if a known object did unexpectedly move, the robot could crash into something (or weld the air as in the earlier example). Planning was also slow. When I typed a command into Handey, telling it to reposition a part, it would often sit and think for 20 minutes (!) before starting to move. Handey was serenely contemplative. No real-world urgency could hasten its meticulous pursuit of the flawless plan. In the eternal tension between perfection and completion, Handey always chose perfection.

BEHAVING

Insects inspired Rod's behavior-based programming approach. Although they possess the tiniest of brains, insects are proficient at navigating all the complexity, uncertainty, and messiness of the real world. They find food, mates, and shelter. When threatened by a hungry predator, no insect sits and ponders its predicament for 20 minutes; it reacts instantly.

An insect seems to construct neither a mathematical model of the world nor a detailed plan. But no model is necessary if, at any moment, the insect can directly sense all the aspects of the world it needs to know. And no plan is required if the state of the world at any instant informs the insect of what it should do next.

A behavior-based program mimics insects by sensing the world continuously and then reacting to what it senses. When programmers implement a behavior-based program they begin by designing a collection of elementary behaviors. These behaviors can be simple, direct things like always-drive-forward or turn-away-from-an-obstacle or turn-toward-the-brightest-light. Each behavior uses information from the robot's sensors to decide what it thinks the robot should do right now. Behaviors often have conflicting ideas about this, and so there's an extra bit of software called an *arbiter*. The arbiter enforces the rules about which behavior gets its way at any moment. The overall behavior of the robot emerges from the interaction of the elementary behaviors, the arbiter, and the dynamic state of the world. That behavior can be surprisingly rich and complex.

Behavior-based programs were the brash, impatient, James Deans of robotics. Urgency was all they knew—a fraction of a second to perceive, a fraction to think, then act. If that didn't work, try something else. A behavior-based program made no guarantees. It couldn't tell if the task it'd been given was easy or hopeless. In either case, it just kept trying.

This usually worked. Despite the absence of guarantees, a well-designed behavior-based program typically succeeded. And, unlike a modeling-planning program, a behavior-based program didn't require an unrealistic static world. If the world changed, the program reacted immediately.

Rod's students constructed robots to demonstrate and explore the new approach. An especially fun pair of robots were built from 1/24-scale

RC cars.[16] Their remote-control components were stripped out and replaced by basic computing hardware and simple sensors. (The sensors used were similar to ones used today to detect your hands and turn on the water in a lavatory sink.) This, plus their behavior-based programs, gave the first robot the ability to chase the second in the pair as they raced about the room at jogging speed! These robots were called Tom and Jerry.[17]

Embodying the modeling-planning paradigm, Shakey, a famous human-sized robot built at Stanford University in 1966,[18] could perceive only certain predefined geometric shapes. It moved about its sterile domain at a snail's pace. To avoid becoming trapped, many robots of Shakey's era needed to keep well away from any object they couldn't recognize. In contrast, behavior-based Genghis,[19] a six-legged walking robot created in Rod's lab, could follow people and explore typical indoor environments. Although not as speedy as Tom and Jerry, Genghis moved continuously and did not hesitate to touch objects. Indeed, it climbed over any low obstacle and turned away from large ones. As an early example of behavior-based robotics, Genghis ultimately went on display at the Smithsonian Air and Space Museum. Rod's grad student Colin Angle (whom we will encounter again soon) built Genghis.

But the robot I found most intriguing of all was one Jon Connell, another of Rod's students, created. Called The Collection Machine, or TCM, it stood apart from other robots because it wasn't merely an academic exercise. This robot performed an actual, useful job—TCM collected empty soda cans from offices.

In the mid-1980s, a task like soda-can collection was considered to be beyond the capabilities of contemporary robotics. To attempt it, conventional wisdom might suggest a carefully built map of the offices, a localization system to pinpoint the robot on the map, a vision system to identify cans amid clutter, a general-purpose arm and gripper to grab the can, and a large mobile base with sufficient power to move the conglomeration around. All that hardware would likely cost tens if not hundreds of thousands of dollars.

But using the behavior-based paradigm and copious ingenuity, Jon Connell accomplished something of audacious scope for a relative pittance. Importantly, TCM had no map. Instead, it just tried to keep moving and avoid bumping into things. It needed no general purpose vision system (which researchers had yet to master). Rather it used a camera and a wedge-shaped laser beam. This combination produced a relatively easy-to-identify signature when the laser stripe encountered a soda can. TCM had an arm and gripper not really suitable for picking up anything other than objects about the size and weight of an empty soda can. Specializing in this way allowed the arm and gripper to be small and lightweight, which in turn minimized the size of the base and the required power.

TCM performed a fiendishly difficult task, doing so with minimum complexity and at low cost. It didn't even have a central computer! Much of the robotic world at the time was focused on complicated "general-purpose" robots that promised much, cost a bundle, and accomplished bupkis, or nearly so. Rod and Jon showed that a heretical approach—do just one thing as simply as possible—could outperform conventional wisdom.[20] That lesson inspired me. Maybe, I thought, there were many other areas where minimal hardware and maximum innovation could accomplish other amazing feats.

ROBOT TALENT SHOW

January is the Independent Activities Period (IAP) at MIT.[21] It's a month with no formal classes but plenty of quirky seminars, workshops, and other activities. Anyone willing to put in the effort, be they student, faculty, or group, is welcome to organize an event. During IAP the AI Lab had a tradition of holding an "AI Olympics."

The Lab's "Olympics" typically consisted of games and athletic contests. But in 1989 Rod's Mobot Group tried something new. They conceived of, not a competition, but more of a common experience where individuals and teams could build robots of their own design. The Mobot Group would facilitate this by supplying sensors, motors, microprocessors, Legos, tools, documentation, and software. Anyone associated with the Lab could use those resources to build whatever robotic contraption they chose. (Today we'd call this a maker event.) At the end of January participants and spectators would gather at the MIT Faculty Club for a Robot Talent Show.

My fellow research staffer Anita Flynn managed Rod's lab and spearheaded the extensive effort needed to organize the event. She explained the reason they'd undertaken the sponsorship. It seems that the Mobot Group frequently fielded questions from lab members asking what robot they would build next and what would it do? So, they thought, why not just let lab folk build their own robots?

Building robots proved immensely popular. Some 60 students, staffers, and others sorted themselves into 20-plus ad hoc teams. Much of the Lab's official research screeched to a halt during the month of January while we makers devoted all our attention to bringing forth our creations. Since I had previous experience constructing and debugging hardware, I volunteered to advise and assist my fellow robot builders.

The idea of creating a robot of my own was very exciting—I just needed a good project. At the time, I lived in a one-bedroom apartment near Central Square, Cambridge. My roommate-less living situation sparked a dilemma: although I valued cleanliness, the act of cleaning held negative appeal. My vacuum was a lonely machine. A robot that could clean the floor of my apartment seemed the ideal solution—cleanliness without cleaning!

As we saw earlier, the notion of building a robot to clean home floors was a popular and frequently attempted endeavor—I'd heard about a few of those tries. So had another team in the AI Olympics who also settled on the robot floor-cleaner idea. They planned to call their machine "Zoom Broom." Thus prompted to consider it, I suddenly found the status quo disappointing and confusing. Why was it, I wondered, that no robot vacuum had ever made the leap from research to retail? Maybe, I'd understand if I tried to build my own.

'TIS A GIFT TO BE SIMPLE

"I'll build the Liberace of robots," I first thought. "My robot will have every bell and whistle I can think of. Its virtuoso performance at the talent show will be the envy of all!" Fortunately, the prudent angel on my other shoulder spoke up. "Your team is just you. You'll have to conceive, design, construct, program, and debug every system and component on that robot." That was daunting given that I had a month to progress from point A, never having built a robot before in my life, to point B, a fully functioning device performing in front of a crowd. Every one of those fancy frills I was imagining would act like a tiny vampire, sucking time and effort away from working on the thing I most wanted the robot to do—clean the floor.

That led to a second thought. "I'll build the Gandhi of robots. I'll strip away everything superfluous leaving only what's necessary to do one task!" Aiming for a more modest robot seemed like a better plan but it begged the question: what exactly does the simplest possible, one-task robot look like? As mentioned earlier, I had plenty of experience building circuits and mechanisms (or at least trying to). My takeaway from many failures and a few successes was this: start with a crystal-clear idea about what the device must do. That's the only way to avoid time-wasting tangents.

In one sense, my robot's task was simple, obvious, and easy to state: Clean the floor. Any human given that instruction would know just what to do. But a robot finds considerable ambiguity packed into those three words. Should the robot clean all types of floors, say tile, low-pile carpet, and shag? Is dealing with the clothes, books, and papers that often cluttered my apartment floor part of the job? And ought the robot recognize and respond to the danger presented by the top of the stairs or other precipice?

We take for granted as commonplace innate human abilities like perception, dexterity, and common sense. But these and other competencies—the product of eons of evolution—are in fact most extraordinary. Such faculties establish the foundation for everything we do from sweeping a floor to flying a spacecraft to the moon.

My robot began at a great disadvantage, because it had no innate abilities at all. For any task I wanted it to perform I'd have to devise a strategy, implement sensors and other hardware, program behaviors, and then test, fail, and try again. (Another lesson from experience is that nothing works the first time.) The Talent Show was a month away and the clock was ticking. A key tactic engineers and others use to accomplish a goal in a short time is to limit the scope. In the case of my Talent Show robot I needed to put some very thoughtful boundaries around exactly what it means to "clean the floor."

My robot's job was to perform a credible act of floor-cleaning before an audience at the Faculty Club. There were no stairs near the Faculty Club floor. So I didn't need sensors to detect drop-offs, and I wouldn't have to program any "cliff avoidance" behaviors. Likewise, I could stipulate a clutter-free performance area. The robot would have to deal only with the "dirt" I brought along for the demo. That let me reclaim any time I'd otherwise have spent working on ways to prevent the robot from ingesting, say socks and shoelaces. But I did want the robot to clean every square inch of floor it could reach and to respond intelligently when it came upon an obstacle. That meant the robot needed a mechanism for picking up dirt from floors (I'll declare shag and other difficult surfaces off limits), and it must not get stuck when it encountered walls, chair legs, and other furniture.

ROUND IS BEAUTIFUL

The not-getting-stuck part is tricky. My boss, Tomás Lozano-Pérez, gained fame and early tenure by inventing a method that lets a robot of any shape or type—mobile or manipulator—find its way around any space no matter how cluttered or complicated.[22] Such Houdini-like ability is exactly what I wanted for my robot. But there's a catch. Tomás' method requires huge amounts of sensory information and major processing power. I had neither. And even if I could afford advanced sensors and a powerful computer, the resulting system would be slow (like Handey) and might take me months to implement.

And I had another thing to worry about. Most robots I'm familiar with (Tom and Jerry for example) possess simple, cheap sensors that give a yes or no answer to the question, "Is there an obstacle nearby in the direction I'm pointing?" That's enough information to let the robot sense an impending collision and turn away before it's too late. But it won't work for my robot. If my robot turns away too soon, it will leave an uncleaned margin around every object in the room. My requirement to clean every square inch of floor means the robot can't give obstacles a wide berth, it has to get as close as possible—effectively, it has to touch. Once in contact with an obstacle the robot must realize that its way is blocked and then choose a different path so it can keep on cleaning.

Knowing when the robot hit something wasn't too difficult. I could build a floating bumper, a spring-supported shell surrounding the robot chassis. Then I could mount electrical switches between the chassis and bumper. When the bumper touched an obstacle, the force would be transferred to one of the switches, closing it. The robot could then compute the direction of the obstacle by noting which switch was closed.

So, suppose that the robot touched an object, and that its sensors let it detect the collision and the approximate direction of the object. What next? It turned out that there was a very simple, reliable strategy to let the robot escape this close encounter: The robot should spin about its center, point itself in a direction that's not blocked, and then drive forward. Escape accomplished. But this works only if the robot is *round*.

Simple geometry can make life easy for a robot or a person. Have you ever tried to carry an unwieldy couch up a narrow, twisty flight of stairs? This situation is vexing partly because, before you try it, you can't be sure whether success is even possible. When, after a tedious and exhausting effort of hefting and twisting, you find yourself blocked halfway to the top, should you give up or try to find a more creative sequence of turns and shifts to bring home the Chesterfield?

If that wasn't hard enough, now imagine moving the couch in complete darkness. The only information available to you is the force you feel when trying to push or twist the couch in some direction and finding that it won't go. This is the same confusing problem a non-round mobile robot faces. When it collides with something, all the robot knows is that what it tried to do didn't work. If it was unable to spin left, should it spin right? Should it first back up? How far? Or does it need to arc backwards to the left or maybe to the right? And what if the next action it tries also fails? And the one after that? Nothing in the robot's sensory data offers any guidance for what it should do next. Metaphorically, it's halfway up the stairs, in the dark, and it's stuck.

What if we replaced the couch we're trying to carry upstairs with a large disco ball? The problem suddenly becomes much simpler! Now there's no need to guess an intricate sequence of twists and turns. If the ball is small enough to fit, then the passage itself will guide it along. If you can't quickly find a way to push it forward, it's because the ball is simply too large. Partygoers on the second floor will have to content themselves with a mirrored sphere of a less awesome size. In the same way, a round robot eliminates the need for all sorts of complexity in sensing and programming because spinning in place always works. For the talent show, I need things to work right away. So, my robot should be round.[23]

And how will I make sure the robot cleans the *entire* floor? I'll rely on chance. It's not very elegant to have my robot just bounce around randomly—rolling the dice to pick a new direction every time it bumps into something—but it works. If I simply let the robot run for long enough it's mathematically guaranteed, ultimately, to visit and therefore clean every point on the floor. Why is it guaranteed? Because the robot is casting honest dice. Roll dice enough times and every possible number will come up. If the numbers represent the paths the robot might take, then eventually it will take every path.

ZOOM BROOM

Graduate students Lukas Ruecker and Tanveer Syeda comprised team Zoom Broom, the other AI Olympics group building a robotic floor cleaner. Lukas and I often encountered each other in the playrooms and hallways of the Lab and that gave us a chance to compare notes on our progress. I'd chosen to construct my robot's chassis from Lego. And, although I loved the fact that this let me build and alter structures quickly, it was frustrating that Lego bricks' studs and tubes forced me to place components where they fit rather than precisely where I wanted them. Lucas and Tanveer had dispensed with Lego altogether, choosing instead to attach motors and sensors to the body of a square-shaped (handle-less) Bissell carpet sweeper.

Lukas and I disputed the wisdom of our respective approaches. I thought a fully integrated, built-from-scratch device simplified the robotics problems. A repurposed bottle brush was the heart of my cleaning mechanism and getting that built proved to be the most difficult aspect of my project. Lukas argued that a modular method was better overall. He and Tanveer spent no time making their cleaning mechanism work—the Bissell sweeper solved that problem. Their effort was directed at robotizing the sweeper by attaching motors and sensors and programming its motion.

It was mostly because I wanted my robot to be round that I was forced to build my own cleaning mechanism. The advantage this gave me was that my robot would be able to get into tight spaces without getting stuck and I could spend minimal time developing my navigation algorithm. But I did have to work much harder on the mechanics. Lukas and Tanveer's robot would clean well but it needed to stay away from tight spaces.

I finally got all the mechanical components of my robot to fit together (I had to abandon purity and use a few non-Lego components) about two days before the Talent Show. So it was fortunate that I didn't need to spend much time programming. But getting to a final design required many iterations. One prototype, I thought, resembled the quixotic vehicles in the *Road Warrior* movies. So I called my robot "Rug Warrior."

A NIGHT AT THE (ROBOT) OPERA

An appreciative and encouraging audience packed the Faculty Club on the evening of the Robot Talent Show. And, even more exciting, a crew from the local PBS station, WGBH-TV, showed up to televise the revolution.[24] Among the many performances that night, a three-foot-long autonomous robot blimp motored above the crowd, a robotic inchworm crossed the stage, and a tiny "downhill skier" attempted a run down a cardboard ramp. A crowd favorite was Henry Minsky's robotic pig pusher. Originally intended to locate and remove trash, Henry decided it would be much more fun to use his robot, based on an RC bulldozer, to chase down and push away a walking, grunting toy pig.

Alas, Lukas's and Tanveer's Zoom Broom had somehow offended the demo gods. When the power came on, rather than triumphantly clean the stage, Zoom Broom just twitched a bit. Lukas was left to describe to the crowd what his team's pride and joy was supposed to have done. Been there.

Rug Warrior was luckier. When my turn came, I crumbled some Styrofoam "dirt" onto the floor and switched on the power. Moving in a square spiral, the robot drove over the dirt and even managed to pick up some of it. And Rug Warrior correctly turned away from obstacles I put in its path. Following my prudent angel's advice paid off.

However, that jubilant moment masked a critical flaw. In my haste to complete Rug Warrior I'd made a mistake and had embedded in my design a misunderstanding of a crucial system. Hidden from view, it would lurk for 13 years. Then, with the cruelest of timing, it would appear, preventing Roomba from cleaning, and confounding our team. But that was a matter for another day.

Participating in the Olympics, building Rug Warrior, and seeing it come to life was some of the greatest fun I'd ever had. My fellow Olympians enjoyed similar experiences. Such delight shouldn't be the purview of the AI Lab alone, I felt. So, later I approached Anita Flynn, whose robot-building experience exceeded my own, and told her my thoughts. Lots of people might like to design and build their own robots if only someone would show them how; I proposed that we collaborate on a book to do just that. Anita said that sounded like a fun idea. The result a couple of years later was *Mobile Robots: Inspiration to Implementation*.[25]

But the Robot Talent Show failed to answer my question, "Why are there no robot floor cleaners?" Rug Warrior wasn't able to displace my home vacuum cleaner as I'd hoped. Lego was not the material of choice when ruggedness and long-term reliability were required. Also, shortcuts I took to finish on time made Rug Warrior awkward as an everyday tool. For example, the dirt-collecting bin was held on with masking tape and the batteries were not rechargeable. Still, it seemed to prove the concept. If one recruited a team of engineers and gave them resources and time, I could see no reason why Rug Warrior or a similar robot couldn't become a real product capable of cleaning real floors.

After the AI Olympics, further development of a robotic floor cleaner felt interesting but not compelling. Working at the Lab was just too much fun to contemplate any other life. The idea of abandoning the sublime pleasures of academia for the crass world of business just so I could build a new product held less appeal than vacuuming my apartment. "No," I thought. "I've found my niche; I'll stay at the Lab forever."

NOTES

1 I believe the space probe I saw was the Explorer 1. This was the US's first successful satellite, launched in February 1958—hence the happiness and excitement.

2 A flyback transformer generated the high voltages the cathode ray tubes found in old TV sets needed. The component's name and function were mysteries to me at the time. I just liked the big sparks it made.

3 I studied a few key books and magazines I'd acquired. I fondly remember getting a subscription to *Scientific American* around 1966 at age 13 (paid for by mowing lawns). Next, I discovered model rocketry. Designing my own rockets was the most fun. Not every model flew as intended but at least none just sat on the ground!

4 I'd been studying plasma physics in the hopes of contributing to fusion as an energy source. But having never managed to get excited about the details, I left without completing my PhD.

5 The William H Bates Linear Accelerator in Middleton, Massachusetts was operated by MIT.

6 The Lab is now known as CSAIL (Computer Science and Artificial Intelligence Laboratory).

7 To be precise, the whiteboard was on the seventh floor; there was a blackboard on the eighth.

8 The Unimate robot, invented by George Devol and promoted by Joseph Engelberger, entered service at a General Motors factory in 1961. The robot consisted of an arm and a gripper. It was used initially to transfer die-cast parts.

9 T. Lozano-Pérez, J. Jones, E. Mazer, and P. O'Donnell, *Handey: A Robot Task Planner*, MIT Press, 1992.

10 In robotics, the pose of an object is the combination of its position and orientation.

11 I mean no disrespect to the world of manipulator robots! Such robots assemble products, weld, paint, and perform many other critically important tasks. They have contributed immeasurably to the world economy. I just found mobile robots—entities that move around the same spaces I occupy—to be more interesting.

12 Mobot is an acronym for MObile roBOT.

13 I followed Rod's group from the sidelines. As a Lab employee, I couldn't have just switched groups even if I'd wanted to.

14 That is to say plans produced by Handey offered the same certain validity as mathematical proofs.

15 "Planning Is Just a Way of Avoiding Figuring Out What to Do Next," MIT AI Lab Working Paper 303, Brooks, R.A., September 1987.

16 Radio-controlled model cars. The scale number means that if the real car was 14 feet long, the model would be 1/24 that length or seven inches.

17 Jonathan Connell, "Creature Design with the Subsumption Architecture." Proceedings of the 1988 SPIE Conference on Mobile Robots, 383–390.

18 Nilsson, N. J., et al., "Application of Intelligent Automata to Reconnaissance," Final Report, Stanford Research Institute, December 1968.

19 Colin Angle, "Genghis, a Six-Legged Autonomous Walking Robot." Thesis, Massachusetts Institute of Technology, 1989.

20 Jonathan H. Connell, "A Colony Architecture for an Artificial Creature." Thesis, Massachusetts Institute of Technology, 1989.

21 The first IAP arrived at MIT in January 1971, preceding me by nine months.
22 In computer science, moving an object of arbitrary shape (including the robot itself) about a cluttered space is known as the "Piano Mover's Problem." Tomás was the first to find a general solution. It's described in his paper: T. Lozano-Pérez, "Spatial Planning: A Configuration Space Approach," in *IEEE Transactions on Computers*, vol. C-32, no. 2, pp. 108–120, February. 1983, doi: 10.1109/TC.1983.1676196.
23 Mathematicians will object that the piano mover's problem as described is six-dimensional while the floor-cleaning robot need deal with only three dimensions (x-position, y-position, and orientation). True, but when we make the piano (couch) a sphere, or the robot a circle we simplify either problem by an order of magnitude or so—which is the point.
24 "Ten O'Clock News; Robot Talent Show," 2/3/1989, WGBH-TV.
25 J. Jones and A. Flynn, *Mobile Robots: Inspiration to Implementation*, 1st edition, A. K. Peters, 1993.

CHAPTER 3

Unfinished Symphonies

THE FIRST GULF WAR raged for several months in 1990 and 1991. Tomahawk missiles were one of the weapons the United States used in that conflict to reverse Iraq's invasion of Kuwait. Each missile cost around $1 million and 288 were launched during the war. Preoccupied by another matter as he watched news of the unfolding tragedy, Tomás Lozano-Pérez couldn't help but think, "For the price of just one of those missiles we could fund research at the lab for a year!" That "other matter" on Tomás' mind was the rapidly approaching end of the initiative that paid the salaries of the AI Lab's research staff. The multi-year grant had run its course, and Lab director Patrick Winston, my boss Tomás, and others desperately sought another source of funds.

Heroic efforts notwithstanding, no new cash was found. So early in 1991 Tomás called me to his office to share the bad news. "You'll be out of a job in September," he said.

Suddenly, I felt like Handey trying to plan in a world where objects wouldn't stay put.

Having discovered my passion for robots I wasn't willing to switch careers; only a job in robotics would do. So I beseeched professors who had their own funding, investigated grants, and brainstormed novel schemes to remain employed. But I could find no way to stay at the Lab. Expulsion from academia and exile to the business world was fate's unappealable verdict.

DENNING MOBILE ROBOTICS

In the early 1990s there were so few robot companies in the United States that I could name every one. (Today, I can't name all the robot companies in Massachusetts, my home state.) But despite the poor odds, providence

32   DOI: 10.1201/9781003540489-3

smiled on me. A member of that rare fraternity of robot firms, Denning Mobile Robotics, was within commuting distance of my house. Even better, Phil Veatch, an acquaintance of mine, worked there.

During the spring I contacted Phil. It turned out that one of Denning's programmers had recently left—potentially creating an opening for me. Phil talked me up to his boss and I got an interview with Bulent Sert, Denning's Director of Engineering. Our chat went well, and later Bulent called to welcome me aboard! Beginning in the summer I started splitting my time between the Lab and Denning, ramping down at the former and up at the latter.

Full-time work started on the first of September when I joined about 20 other Denning employees in an impressive two-story building at 21 Concord Street, Wilmington, MA. The entrance featured an atrium dominated by a crafted-in-wood version of the Denning logo, a sort of strange-looking mobile robot that I never quite figured out. Pieces of robots in various states of assembly could be found throughout the building. On the second floor was a conference room and a large open area—great for testing robots I thought. The first floor housed a large assembly shop, Bulent's office, and the company's main computer. The building was mostly surrounded by woods and a favorite lunchtime activity among the engineers was to stroll down to a nearby bridge. If we were lucky, the resident turtle might put in an appearance.

Whereas the currency of the AI Lab had been ideas at Denning it was business—expressed in specifications, budgets, and contracts. The Denning ethos was different from the Lab in another, less comfortable way. At the Lab we'd ask, "What do the results of the experiment show?" Here the question was, "What does the president want?" This new world was foreign to me but, for the privilege of working on robots, I'd try to accommodate.

When I joined the company, Denning had been in business for eight years. Their flagship product was the Denning Sentry robot. Sentry's purpose was to be "the eyes and ears of your security guard." Effectively, that meant Sentry autonomously carted a collection of sensors around a warehouse or office building, transmitting the information they gathered back to the security post. The sensor suite could include a microphone, video camera, smoke detector, motion sensor, and so on. The robot was cylindrical, just over four feet tall, and weighed 465 pounds. Back then it cost about $75,000.

Denning also made research robots for universities based on the Sentry platform. The company had developed an automated camera pedestal for TV stations and, in partnership with Windsor Industries (a manufacturer of commercial cleaning equipment), work was in progress on an autonomous autoscrubber called RoboScrub. You've undoubtedly glimpsed a worker piloting a manual autoscrubber as it cleaned the hallways of an office building or school late at night. Shortly we'll discover another robotic autoscrubber lurking in Roomba's pre-history.

Denning's CEO and President R. Warren "Bob" George excelled at coaxing funds away from investors but the company's recent performance hadn't been living up to the rosy projections of corporate presentations. Current sales could not sustain the business. To "extend the runway" Denning laid off four technicians the week I started working full time. That was unsettling but, I thought, the state of Denning's business might make them receptive to going in a new direction.

I made my case to Bulent. "All of Denning's robots are big and expensive," I said. "But there's a new theory that lets small, inexpensive robots do useful things. It'll be easier to sell robots if they cost less."

"Bob George likes big robots," Bulent said.

To skirt the impasse, I proposed building a little robot floor cleaner. Founded on behavior-based principles it would showcase the advantages of the new paradigm. I felt certain that Denning's CEO would embrace the revolution as soon as he saw the power of the new idea. After a few days of negotiation Bulent and I got to yes. He told me that I could work on the robot as a back-burner project. And I could even call on Denning's mechanical engineer, Jack Shimek, to help me. Things were looking up!

Previously, I'd had some success convincing Jack that small robots could do big things, so it was fun when we began working together. Jack offered to bend the sheet metal for the new robot and to find the motors we would use to power the wheels and the cleaning brush. I would do the rest: architect the design, assemble the electronics, and program the behaviors. I still had a microprocessor board left over from the Robot Talent Show that would become the brain. It was Jack's idea to call our creation "RoboBroom."

Being on the back burner, RoboBroom came together slowly over the course of maybe three or four months. The new robot partly followed Rug Warrior's blueprint: two drive wheels, one on either side, plus a roller for balance, and a floating bumper to detect collisions. But, despite the analysis I did for Rug Warrior, we made RoboBroom square-shaped.

An inevitability in mobile robotics is the tension between design choices that simplify construction and those that simplify programming. It's undeniably easier to pack components into a square robot than a round one. It's also true that the program enabling a round robot to avoid getting stuck is more reliable and easier to write than the program for a square robot. This time, simplifying the mechanics won, but it nevertheless meant we'd have to be careful with our demonstration to management. To ensure that RoboBroom performed well we'd need a benign venue with no narrow spaces or difficult combinations of objects.

Around the first of December 1991 we were ready. Jack and I set up the demo in that large open area on the second floor. We placed a few obstacles for RoboBroom to bump into and scattered simulated dirt for it to clean. The whole company showed up to watch—not unlike the earlier Robot Talent Show.

RoboBroom worked as designed, moving about the floor, pirouetting away from obstacles, and picking up the dirt. The crowd applauded enthusiastically! Jack and I were very encouraged, and we eagerly anticipated the next step, a business plan to turn RoboBroom into a product.

A few days later, in the early afternoon on December 10, my colleague Austin told me, "Bulent wants to see you in his office."

I walked down to Bulent's office on the first floor and stepped inside. On his desk, I was surprised to see a check made out to me. Bulent said, "We've decided to … Bob George has taken the decision to lay off you and Jack. Here's your final paycheck." Bulent offered some explanations about the state of the business, the need to conserve cash, and so on. But he ended with, "Bob thinks you and Jack are just playing with toys."

My stint at Denning was over almost before it began.[1]

IS ROBOTICS

This was the second time in less than a year that I'd been laid off from a robotics job—it put me in a bind. The AI Lab let me know about my impending unemployment six months before the fact. Denning's notice was more like six minutes. By this point I was established—my wife and I had a toddler, a mortgage, and plenty of bills to pay. A job—even a nonrobotics job—was a necessity. I began my search the evening I was laid off.

I told everyone I could think of that I was looking for a job. I scoured newspaper listings, made calls, and left messages. Remarkably, in only a week or two I uncovered an opportunity more exciting than anything I'd dared to hope for. A year earlier, Rod Brooks, Colin Angle, his wife Jill

Angle, and Helen Greiner had joined forces to start a new robot company. All had been associated with the AI Lab, and they were now determined to commercialize robots based on Rod's new behavior-based robot control paradigm.

They'd gotten assistance in this endeavor from a California firm called ISX. Thus inspired, Rod, Colin, Jill, and Helen decided to call their new company IS Robotics. (I believe their subsequent accomplishments atone for the awkwardness of the name.)

As quickly as possible, I got in touch with Rod. "I'm looking for a job," I said.

"We don't have any money," Rod told me.

But they were on the trail. The team had already developed hardware useful for the sort of robots they planned to build. One element of this was a more capable version of the Robot Talent Show microprocessor board; another was a standard for connecting the onboard computer with other robot components. When I called in January 1992, Rod and Colin were negotiating a contract with a Japanese company that wanted IS Robotics (ISR) to build a demonstration robot for them.

The next few weeks were stressful. The folks at ISR practiced robotics exactly the way I believed robotics should be done. I very much wanted to work there but because that job was in the same state as Schrödinger's cat, I had to keep looking for and applying for other jobs for which the wave function had already collapsed.[2] (Physics major, remember?)

In February 1992, IS Robotics got the contract, and I got the job.

Up to this point ISR had been billeted in Colin and Jill's apartment; the new contract enabled the company to rent an actual office. But financial realities made it necessary to continue operating on a shoestring, even after the first check arrived. None of us received market-rate salaries, and where Denning's nerve center had been a notch below opulent, IS Robotics' new home was a notch below basic.

ISR's global headquarters consisted of a space on the second floor of 238 Broadway in Cambridge. We had a conference room, a room for electronics and mechanical assembly, and an open area for our desks. Walking across the floor could be unsettling as the uneven, creaking, not-quite firm floor inspired no confidence. A dropped ball would roll to one corner. One day, as it was starting to rain Colin said, "Come and help me spread plastic." "What?" I asked. Colin explained that whenever it rained, flecks of plaster and paint would slough off the walls and get into the electronic assemblies on the table below.

NO DATE FOR THE DANCE

I told my new colleagues the tale of woe that was Denning Mobile Robotics. Predictably, they all loved the idea of a small robotic floor cleaner. But as appealing as that project might be, it was clear that we couldn't attempt it. The problem was resources. Even the simplest consumer product can require millions of dollars to design, test, and market. ISR entered the business world bankrolled by dreams and audacity, not venture capital. As such we lived invoice-to-paycheck—deliver a robot, get the check, make payroll, repeat monthly. Our circumstances consigned expensive undertakings like developing a consumer product to our wish list.

So, we looked for another way. "Maybe someone will partner with us," Colin suggested. The obvious candidate was Bissell. Their products had provided the guts of both the AI Olympics' Zoom Broom and the recently spurned Denning RoboBroom. "I'll see if they're interested," I told the group. With great care I composed a letter. Hoping to evoke excitement, I suggested how Bissell's traditional products might be reimagined and reinvigorated with cutting-edge technology. Bissell could get a jump on competitors and grow their venerable business by offering a convenience no one else could match. In keeping with those primitive times, I put a stamp on the envelope and dropped it in the mail.

A few days later we got a letter in reply! But it wasn't the yes or no answer we expected. Rather, Bissell's missive informed us that they had received our letter but would not read it until we agreed to certain legal protections for them. They asked for a nondisclosure agreement and maybe another document or two. These we sent.

We heard nothing for a couple of weeks, so I called the contact number Bissell had provided. A pleasant gentleman answered. "What do you think about our idea for a little robot that cleans floors?" I asked.

"Well," he said.

> It's intriguing and we talked it over in the office for a while. The problem is your robot is still a carpet sweeper, and nobody will pay more than $20 or $30 for a carpet sweeper. The robot would cost more than that so, Bissell isn't interested.

Bissell's reasoning was as concise as it was crushing.

We tried one more time. Rod seemed to have contacts all over the world and one was with a Korean company called GoldStar (since renamed LG). We had already engaged GoldStar on another matter, so we pitched the idea of a floor-cleaning robot to them as well. This time we got as far as making some drawings and describing how the robot might operate. But after some initial interest, nothing came of our efforts.[3]

We could find no deep-pocketed partner willing to finance our revolution. So, the idea of a robot floor cleaner was shoved to the back burner—and there it would sit neglected for nearly eight years. To achieve the fame and fortune for which we were destined, we'd have to hitch our wagon to some other star. Or so it seemed.

NOTES

1 It surprised my wife when I came home early the day I was let go. For years afterward, she always greeted me with alarm on any workday when I arrived home earlier than expected.

2 A wave function describes a particle's or system's many possible states before observation. Once observed, the wave function is said to collapse, settling on a single reality. Physicist Erwin Schrödinger proposed a thought experiment in 1935 that put a cat in a superposition where it was both alive and dead. This was to suggest the absurdity of one interpretation of quantum mechanics. But it's now widely accepted.

3 It turned out that GoldStar was more interested than we were aware. Around this time they were filing some patents for a home floor-cleaning robot of their own, e.g., US5353224A.

CHAPTER 4

Power

DOI: 10.1201/9781003540489-4

DILEMMA

Anyone who wants to build a robot vacuum must address two questions. First, how will you accomplish the cleaning? Second, how will you power the robot? Both questions have obvious answers. Unfortunately the answers are contradictory.

Since the early 1900s the best way to clean a carpet has been well understood.[1] Use a motor to spin a brush made of stiff bristles. The bristles pummel the carpet, dislodging particles of dirt. A strong vacuum, established by another motor, then sucks the dirt into a canister or disposable bag. The system works so well that practically every household owns a manual vacuum cleaner that operates in exactly this way. At first glance, it seems like a no-brainer that a robot vacuum cleaner should adopt this long-proven technology.

For power, most manual vacuums (at least in the time before Roomba) used a long cord plugged into an electrical outlet. But copying that solution would cause headaches for a robot. The robot might run over and then ingest the cord, or wrap the cord around furniture, immobilizing itself. So, obviously, the robot should ditch the cord in favor of a rechargeable battery—the same solution employed to good effect by millions of portable appliances.

So far, so good—we'll build our floor-cleaning robot on the foundation of two proven technologies. The dilemma emerges when we do the math.

How big must our robot's battery be? The complete calculation is tricky, but the answer will depend principally on three things: (1) how rapidly our robot consumes power; (2) how long the robot must run before it recharges;

and (3) what type of chemistry our battery uses. Chemistry decides a critical battery parameter called *energy density*. Energy density tells us the maximum amount of energy that can be stuffed into a battery of a certain weight and size.

What are reasonable numbers for the terms that go into our battery computation? Let's use a typical manual vacuum as our power role model. These appliances consume about 1200 watts on average, so let's pick that number. Cleaning for say an hour before recharging seems reasonable. And finally, let's give our robot an inexpensive, standard lead-acid battery—the kind of battery that's occupied an honored place under the hood of nearly every automobile for the past 100 years.

Lead-acid batteries vary somewhat, but we can expect our battery to provide an energy density of at least 25 watt-hours per kilogram of mass. That gives us 1200 watts divided by 25 watt-hours per kilogram times one hour which equals 48 kilograms. At 2.2 pounds per kilogram that means our robot's battery will weigh 106 pounds.

It seems like something must be wrong. One hundred pounds is really heavy, and we haven't even accounted for the rest of the robot yet. Most electromechanical devices weigh more than the battery that powers them—the first Roomba weighed 2.7 times as much as its battery. If we match that ratio, then the robot we're designing will tip the scales at about 392 pounds.[2] It might be problematic to have such a bruiser wandering about the house. And if size is commensurate with weight, our machine will surely be too big to clean under the bed or sofa. What can we do?

We have three choices: we can reduce the cleaning power, reduce the runtime, or change the battery chemistry. Maybe we don't really need 1200 watts; that's the average but many vacuums use less. So let's cut the power in half to 600 watts. We might also reduce the runtime, but that's a little chancy. Our robot may not move as fast or clean as efficiently as a person, so, to be on the safe side, let's leave the runtime at one hour. But depending on the decade in which we design our robot, we may be able to choose a better battery chemistry.

Early robot vacuum builders were stuck with low energy-density batteries. But in 1989 a new rechargeable chemistry known as nickel-metal-hydride (NiMH) was introduced. The first such batteries offered at least twice the energy density of lead-acid batteries. Choosing this item, we've now improved two of the terms in our equation by 50 percent each. This means that our now 600-watt, one-hour-runtime, NiMH-powered robot will have a 26-pound battery, and the robot will weigh in at 98 pounds.

That's much better—but still uncomfortable. No one will want to lug such a machine upstairs to clean the second floor—especially not grandma and grandpa. One solution is to simply put off designing our robot a few years to give researchers time to improve battery technology.[3] But that's unsatisfying for anyone who wants the future to start now. Another approach is to further depower the vacuum. But that's also awkward. By their nature standard vacuums need considerable power—give them less and they will clean less well. If only there was some less power-hungry way to clean!

This example illuminates a perplexingly prevalent issue in robotics. We want a robot to perform some task. Manual equipment accomplishes that task using an effective, long-proven method. But some unrecognized constraint prevents a robot from using that same method.

The only way forward is to rethink the task from the ground up. We cannot simply start with the standard solution and then robotize it. Because robots have different strengths and weaknesses from people, we must seek solutions that leverage robots' strengths while mitigating their weaknesses.

CARPET SWEEPER

When I was a kid, my mother had an old carpet sweeper. It was purely mechanical, made of wood, and looked like the Bissell sweepers pictured in vintage catalogs from the late 1800s. (Melville Bissell patented[4] his marvel in 1876.) Our sweeper was supported by four wheels that connected to a cylindrical bristle brush. Pushing the sweeper in either direction made the brush spin and, via some magic that eluded me then, this caused dirt to leave the floor and collect in a pair of bins on either side of the brush.

The sweeper was a breeze to use. Even at a young age I had no trouble pushing it around. And it did a surprisingly good job of cleaning—especially if used repeatedly on the same spot.

We also had a vacuum cleaner,[5] a brash machine that pierced the room with a deafening shriek as it ran. Switch it on and our family's feline member would instantly catapult herself from the room. But the vacuum cleaner made quick work of dirt. It possessed the raw power to extract even detritus that had settled deep into carpet fibers. To get good results using only the carpet sweeper my mother would probably have needed to wield it every day, denying the surface dirt an opportunity to sink in. By opting for the vacuum cleaner, Mom saved herself time and effort.

The salient difference between the two cleaning mechanisms is power. All of the power to operate the sweeper comes from the person pushing it.

A multi-megawatt power plant typically supplies the vacuum. And, as we saw above, while the sweeper sips power, the vacuum devours it. If, instead of plugging into a wall outlet, you used your muscles to source all the power the vacuum needs by, say, pedaling a generator, you'd find yourself exerting as much effort as a sprinter running a 100-meter dash in 10 seconds— not an exaggeration! But you couldn't stop after 10 seconds, you'd need to keep pedaling just as hard for 15 minutes or so until the floor was clean.

The manual vacuum cleaner is time efficient but power-profligate. My childhood experience showed me that it's possible to clean a floor well while consuming vastly less power than a typical vacuum cleaner.

Power was one of the critical issues I wrestled with while designing Rug Warrior and RoboBroom. The standard vacuum-cleaning solution wasn't available to the robot because it had to be powered by a battery. I felt I'd found the perfect solution in the carpet sweeper. The superb power efficiency of that mechanism gave me a decent run-time with a modest battery. To achieve power efficiency I traded away time efficiency. Lower time-efficiency meant that my carpet sweeper-based robots would need to run longer to achieve the same level of cleanliness as a manual vacuum. But so what? The robot has nothing else to do—it lives to clean!

Later we'll collide with a non-technical problem that didn't occur to me when I decided that the carpet sweeper was Melville Bissell's gift to robotics. The issue in a nutshell: Everyone knew a traditional vacuum-cleaning mechanism was the best way to clean floors, but no one understood the just-explained subtlety that prevented robots from using it.

NOTES

1 Carroll Gantz, *The Vacuum Cleaner: A History*, McFarland & Company, 2012.
2 That's even beefier than the average offensive lineman in the NFL.
3 That was and is happening. By the year 2000, new lithium-based batteries with energy densities of (at the time) maybe 150 watt-hours per kilogram came to market, although they were frightfully expensive at first. But spurred on by the needs of electric cars, battery prices have come down and energy density has gone up. If we'd waited another 20 years before inventing Roomba, power would have been a less tricky problem.
4 US patent number 182346. Melville Bissell wasn't the first inventor to build a carpet sweeper. Earlier versions go back as far as 1811. But his implementations were the most successful.
5 It was an Air-Way Sanitizer model 66, I have since learned. When my family got a new vacuum I disassembled the Air-Way so I could use its motor and other components for projects and experiments.

CHAPTER 5

Late Bloomer

I N THE EARLY YEARS of IS Robotics, our financial situation fell far short of our aspirations. Rather than leading the way to a bold new robotic future we barely kept the lights on. Still, we felt we had received a calling. Fate had chosen us to unlock the robot revolution using the master key of behavior-based programming. Looking ahead we saw legions of robots emerging like 17-year cicadas to fill every feasible robotic niche. We imagined that each one of those robots would sport an ISR logo. Our robot vacuum cleaner pitches to Bissell and GoldStar might have fallen short but, never mind, a world of tantalizing robot possibilities remained.

We began our corporate quest driven by faith in our paradigm, bountiful enthusiasm, and considerable technical chops. But little else. Our challenge was that we had no investors, little practical business experience, and, worst of all, no viable strategy for transforming our zeal for robots into cash. We needed a business plan.

To find a plan we searched high and low. Literally, from outer space to miles beneath the earth. A couple of the business plans we pursued were actual moonshots. One idea was that we would design a rover able to function on the lunar surface. Next, we'd find a way to get it to the moon. Then, for a fee, we'd let users on earth drive the robot about the moon. Either that or sell the movie rights. Hanging on the wall of our Broadway Street office was an image of a white, tracked robot proudly displaying the ISR name, nestled amid the craters and ridges on the moon. We did not lack for bold ideas. Ultimately, though, the rover-on-the-moon plan withered before the harsh reality of space transportation, not to mention harsh realities of many other sorts. But we were just getting started.

DOI: 10.1201/9781003540489-5

Another, more temperate idea, was to build robots we could sell to others to facilitate their research in robotics. We could claim some expertise here as we had just emerged from a robotics research lab where we'd developed several widely acclaimed robots. Our line of research robots included Genghis-II and R-2. Genghis-II was based on one of those acclaimed robots, Genghis, the hexapod, insect-like walking robot that Colin had built while a PhD student in Rod's Mobot Group at the AI Lab.

The R-2 robot we developed from scratch. That robot was roughly octagonal, about a foot in diameter and a foot tall. It had two powered wheels on either side for propulsion and a ball caster in the back for balance. R-2's coolest feature was a two-fingered, three-and-a-half-inch long gripper mounted on the robot's front. The fingers could open and close and move up and down. The integrated gripper let R-2 pick up and transport small objects.

Our finances were precarious; each month began with insufficient funds to meet the end-of-month payroll. This stranded us in permanent crisis mode. The only way to get cash was to deliver robots. That made shipping robots our paramount priority.

Many high-tech companies focus on design—they outsource the actual assembly of their products to specialists, contract manufacturers. But we could afford no such luxury. Instead, we hand crafted every robot ourselves. Our connection to the MIT AI Lab enabled us to use the tools in the machine shop there, thus sparing us the expense of purchasing our own.

As deadlines approached, we worked long hours. Normal companies call their carrier to arrange package pickups when they have something to ship. But we never did that. That's because such pickups must be scheduled a few hours before the flight that ferries the packages out of town departs from the airport. Those were critical hours we needed to finish building the robot. To reclaim them, we invariably drove our packed-for-shipping robots directly to the FedEx office at Logan Airport at about 9:00 pm. This was not a practice the FedEx folks encouraged, but somehow Colin always managed to talk our robot onto the airplane.

Our frantic shipping schedule often left us with too little time to test the final product. Routinely we would install the last component, turn the power on, and then check the robot for smoke. If no wisps of burning electrical components appeared, we put the robot in a box and carted it to the airport. I'm not proud of that practice; predictably customers sometimes unboxed robots that didn't work. But at the time the stark choice seemed to be either skimp on quality control or shut down the company.

A less brutal plan, when available, was to get paid first and then build the robot. Contracts for robots designed on commission often included provisions for partial payment in advance. Under such an agreement we created a robot for the Sony Wonder Museum in New York City. It was a walking, camera-toting hexapod called Hermes. Helen and I installed a trio of Hermes robots in the museum. Visitors could remotely control each robot as it strolled about ersatz Martian terrain. Only if guided to the far side of a rock mound did the onboard camera reveal a simulated Martian winking at the user from a crevice.

A second custom robot was designed to demonstrate an idea proposed by a Japanese inventor. The robot, fitted with a large ultrasonic panel was intended to smooth freshly poured concrete. We would build pretty much any sort of mobile robot a customer was willing to pay for.

Advanced funds to get the project started usually accompanied robots built in fulfilment of government grants or competitions. Examples include Ursula and later Ariel. Both robots were designed to walk under water where they would locate and destroy mines planted in the surf zone. Another robot, Fetch, was built to retrieve and render safe unexploded munitions on a battlefield. Working on government projects was sufficiently financially rewarding that we opted to become official government contractors. Government contracting obligated us to conform to a plethora of new accounting and security rules and practices.

Over time we worked on many robots. Grendel (our second moonshot robot) rode in a test rocket, was sprung out after touchdown, and explored the landing area. Our Roams robot conducted reconnaissance over difficult terrain. Robots Remus, Romulus, and Piper climbed pipes for inspection purposes. Our Pebbles robot looked for hazardous material. A robot named "It," an early social robot, interacted with people and, excitingly, appeared on the cover of *National Geographic* magazine. Our thin, 30-foot-long robot called BoreRat[1] descended the vertical, two-inch diameter pipe of an oil well hundreds of feet into the earth—and returned!

Working on all these robots was tremendous fun! And every new robot renewed the dream that this would be The One: The must-have robot that would ignite the revolution and transform ISR from scrappy contender into corporate champion.

In the mid-1990s, the company secured an invitation to visit the office of Representative George Brown, congressman from California and chairman of the House Committee on Science and Technology. Colin, Jill, Helen, and I, traveled to Washington, DC and were met outside a House office building by one of Brown's staffers. He did not escort us to our destination through the front entrance as I expected but instead guided us along a circuitous path across parking lots, up and down steps, and through an obscure door into the building. Somehow, we, along with our robot, managed to arrive inside the congressional office building without being scrutinized by security.

We met with Congressman Brown's staff, talked up robotics to them, and showed off our Genghis walking robot. As we were about to head out, a higher-level member of Brown's staff, Dr. Winston Tao, returning from a meeting, noticed our demonstration in the hallway. Taking an immediate interest, Winston asked us if we could delay our departure for a bit; then he disappeared. A few minutes later he returned, Congressman Brown in tow. We restarted our demo and Genghis promptly clambered over the Congressman's feet. Meeting a congressman was fun and a great honor. But the event crucial to Roomba's future would turn out to be our chance encounter with Winston Tao.

From a PR point of view, robots are sexy. The sweeping appeal (so to speak) of robots brought us considerable media attention. Actor Alan Alda visited our office to film an episode of the PBS TV show *Scientific American Frontiers*.[2] (I got his autograph … for my wife.) Another hero of mine, famed science reporter Robert Krulwich, came to our office to do a story. Robert Weiss, producer of *Naked Gun 2½*, showed up for yet another meeting. And numerous other luminaries dropped by. All the press enticed throngs of hopeful jobseekers to knock on our door. Potential business partners showed up too.

Although our hockey stick moment—when robots suddenly transformed to riches—stubbornly refused to arrive, ISR's fiscal position gradually strengthened, and our roster grew. Business skills improved as we learned the game and became more adept at winning grants and contracts. We left our crumbling redoubt in Cambridge for an above-the-shops fortress at a Somerville shopping mall. And we adopted a less problematic name: IS Robotics became iRobot.

It's unlikely anyone would have deliberately chosen the path we followed. But after several years of wandering in the robotic wilderness we had accumulated some key advantages in business, technology, and public relations that would greatly benefit us when we took on Roomba. And we were about to learn the final crucial lessons we needed. They came in the form of two projects: My Real Baby and Clean.

NOTES

1 US Patent: US5947213A.

2 *Scientific American Frontiers*, "Robots Alive!", Season 7, Episode 5, April 9, 1997.

CHAPTER 6

My Real Baby and Clean

T AMAGOTCHIS WERE PALM-SIZED DIGITAL "pets" introduced to the world in the late 1990s. The devices simulated an alien creature that— once hatched from an egg—required patient nurturing from its human owner in order to grow and develop. Neglecting their care could cause Tamagotchis to became antisocial rogues or simply die.

To indicate their need for food, play, or cleaning, Tamagotchis beeped. They beeped incessantly. Owners obsessed over tending their pet, often checking on them between beeps. But too much attention could "over stress" the Tamagotchi and create the same problems as too little attention. (Over stressing their owners was of no concern to Tamagotchis.) The constant distraction they caused earned the devices expulsion from most schools. Perversely, Tamagotchis became the biggest toy fad of the time.

The moral of the Tamagotchi tale was not lost on iRobot. If kids were enthralled by what was basically an LCD screen, a beeper, and a couple of buttons in a tiny box, then surely something more lifelike and responsive would be even more appealing. A toy based on that concept wouldn't exactly be a robot, but it had robotic elements, and, importantly, it might become the next big thing. We also felt that the toy fitted naturally into our behavior-based programming approach. Through a connection, Colin approached the toy company Hasbro with a new concept. We pitched: "A doll that knows how you're playing with it."

Hasbro was intrigued and ultimately allocated funds for our new project. Because of our earlier social robot "It" the project was initially called Bit for "Baby It." Later, Hasbro rechristened Bit "My Real Baby."

DOI: 10.1201/9781003540489-6

The conceit behind My Real Baby was that sensors and software could make a doll more interactive and therefore more fun to play with. Instrumented and programmed, the doll could tell if the child was caressing it or ignoring it, tossing it into the air, or dangling it upside down. Then it could react appropriately. My Real Baby's interaction with its owner was confined to facial expressions and sounds—motors for moving its arms or legs would not be present. This choice guaranteed that no amount of malware, malfunction, or neglect could make My Real Baby turn rogue. It could not transform into Hollywood's Chucky.[1]

PLASTIC AND THE EFFICIENCY TRAP

It seemed an ideal partnership: we at iRobot knew sensors and software, Hasbro knew toys and markets. But a clash of perspectives soon emerged. iRobot engineers built a concept doll with an expressive face that could display a range of emotions. Our earlier It robot had used several motors to drive its expressions, and the first iteration of My Real Baby incorporated five motors. Hasbro was appalled. A representative told us, "Toys cost $20. You get *one* motor!" That was perplexing; how could we make the doll smile and frown, open and close its eyes, suck on a bottle, and so on with just one motor? It turned out there was a way.

At the AI Lab and during the early days at iRobot, we built our machines primarily of metal. A robot was composed of many mechanical parts, and we shaped each one using conventional machine tools—lathe, drill press, milling machine, and so on. The more complex the part, the more time an operator needed to fabricate it. Because time is money, the cost of any mechanism was proportional to its complexity. This was the metric we all understood.

Toys inhabit an alternate reality, metaphorically. Typically, the components of modern toys are not machined from metal, but rather molded in plastic. The process required first cutting a mold from a solid block of steel. The mold can be very complicated with elaborately shaped cavities and bits that move in and out; it may cost tens of thousands of dollars. Second, injecting molten plastic into the mold at a pressure of tons per square inch. Third, opening the mold and ejecting the part. Whether that part is blob-like in its simplicity or fiendishly intricate—after the mold is paid for—the cost of the part equals the cost of plastic it's made from. Compared to a machined part, it seemed to us, a plastic part was free!

For My Real Baby to get by with just one motor all we needed to do was design a deliciously intricate mechanism to interpose between the motor and the doll's face. (Our mechanical engineers salivated at the challenge.) That mechanism could have any number of gears and cranks and cams and sliders—regardless of the complexity, its cost would be low.

Breaking the complex-equals-expensive equation was new and exciting for us. And Hasbro had one more lesson. The single motor we were allowed could cost much less than we were used to paying.

Deeply entrenched in our thinking was the sense that, because robots use batteries, robot motors must be highly efficient. An efficient motor enables a robot to run a long time before needing to replace or recharge its batteries. Engineers love efficient motors and have been known to wax rhapsodic over obscure concepts like high stall torque, minuscule no-load current, and low cogging.

But good motors are expensive. A single high efficiency motor might easily cost more than the target price of the toy—maybe many times more. Sadly, such triumphs of the motor maker's art have no place in a toy. Less efficient motors, we learned, are made by the millions for pennies. For the disadvantage of having a battery charge last, say two hours rather than four, the price of the toy becomes affordable. We learned one more secret about toys—most toys go to their graves (or rather, the yard sale) without ever having their original batteries changed. This despite their inefficient motors.

My Real Baby was launched in time for Christmas in 2000. Hasbro produced TV commercials, placed the robot in many stores, and did a good bit of marketing. Despite these efforts, only a few thousand units were sold; the Tamagotchis tsunami did not recur. Maybe at $95 My Real Baby was just too expensive. Or maybe the marketing worked against us. One advertisement disclaimed, "Doll's face may make mechanical noises." (It was quite loud.) Somewhere there was a disconnect. My Real Baby didn't live up to our dreams.

However, two crucial takeaways from My Real Baby—that plastic is nearly free and that motors can be cheap—were both essential to Roomba. Earlier, when I had computed a retail price for Denning's RoboBroom and for the cleaning robot we pitched to GoldStar, I arrived at a figure that, today, would exceed $1000. I'm not very good at this sort of exercise—given our traditional preference for metal parts and efficient motors, the price might well have been even higher. Had we attempted to market a cleaning robot without learning the key lessons Hasbro imparted, Roomba would likely have been too expensive to succeed.

A ROBOT FOR SC JOHNSON

In 1996, SC Johnson (SCJ), the multi-billion-dollar, century-old, consumer products company, noticed something about the commercial cleaning industry they served. No one was happy.[2] Many managers at large department stores and similar buildings were unhappy with the contractors they hired to clean their facilities. Contractors, in turn, were dissatisfied with the managers who employed them and with their own workers. And workers were unhappy with their employers. Managers perceived contractors as overcharging and underperforming, while contractors underbid jobs to avoid rejection by managers who demanded unrealistic rates. Low pay and inconvenient hours often led workers to quit without notice, leaving contractors with too few workers to staff a given job.

To SCJ, all that dissatisfaction screamed opportunity! Were they to enter the contract cleaning market, they felt they could improve everyone's lot. They would accomplish this by significantly increasing productivity. If SCJ could provide the new cleaning crews they would hire with better tools, they would make workers lives easier and enable them to complete their tasks more quickly. That would let SCJ charge less for the same work and accomplish more cleaning in the same time, all without asking workers to work harder. High productivity would let them treat their customers better and would result in a large market share for SCJ. One of the most time-consuming aspects of the job was the hours typically spent sweeping, scrubbing, and burnishing the floors. If they could get a robot to do that, the economics of their new venture would look promising.

SC Johnson scoured the marketplace for an autonomous cleaning machine that would enable their plan. They found no suitable products. The autoscrubber Windsor Industries and Denning Mobile Robotics had tried to build never became viable. From another manufacturer a product called RoboKent[3] was available, but its rudimentary navigation scheme wasn't powerful enough to fulfill SCJ's vision. After much diligence, SCJ arrived at iRobot's doorstep. They wanted to know if we could help them revolutionize the contract cleaning industry.

Revolution was our calling card! (Or at least we intended it to be.) So, a flurry of meetings and discussions commenced. These required exchanging visits between iRobot's headquarters above the shops at the Twin City Plaza in Somerville, Massachusetts and SCJ's incomparably more prestigious nerve center in Racine, Wisconsin. Finally, we struck a deal. iRobot would develop a robotic autoscrubber for SCJ. The new machine would sweep, scrub, and burnish the floors of department stores, supermarkets, schools, and office buildings. It would do the job just as well as a human-piloted autoscrubber—working as fast as its manual counterpart and it would avoid damaging anything or anyone in the store. We named the new project "Clean" (reserving our creative juices for designing the robot).

At the end of our deal-closing meeting with SCJ executives I flippantly remarked, "That was easy, we should have asked for twice as much!" Some nervous laughter followed. At the time, I believed I was joking.

PAUL SANDIN

Paul Sandin had an ear for music and rhythm. Growing up, his family figured him for a musician. But like me, Paul was late finding his passion and music wasn't it. He began college a couple of times but felt unmotivated and dropped out. While untethered from academia, Paul worked in a boatyard and at several small theaters. In the theater Paul was often called upon to do carpentry projects and construct mechanisms that performed stage magic. Paul discovered he had a talent for such things.

Prompted by a growing interest in technology, Paul decided to try going back to school at a technical college. This time things went much more smoothly. Through his classes and independent investigation Paul discovered robots. These intriguing machines seemed to combine all the aspects of technology he liked best. Paul set his sights on a career in robotics.

After graduation Paul landed at Redzone Robotics in Pittsburgh. There he applied his talents redesigning a specialized bulldozer intended for nuclear waste remediation (it had to be able to fold up in order to traverse narrow passages). Redzone engineers had prototyped the device using a complex combination of six *hydraulic actuators*.[4] Paul found a way to streamline the design using only a single actuator. Simplicity is in the DNA of every practical roboticist.

In late 1996 Paul and his manager experienced "creative differences" over the design of a robot and Paul was fired. Not long afterward, Paul interviewed at iRobot. We were impressed. At that time we had just begun talking with SC Johnson about the Clean project. Paul spent several months waiting anxiously and taking temporary jobs before we finalized the contract and offered him a position.

CLEAN START

The Clean project excited me. Serendipity had brought us a chance to spark a new industry (commercial robotic floor cleaning) by being first to deploy a fully capable robotic autoscrubber. The concept we'd agreed to develop was big and complex—it would take a big team to pull it off.

Paul Sandin was an early hire for that new team, another was our project director, Lee Sword. Lee came to iRobot fresh from the Jet Propulsion Laboratory in California where he'd played an important role building Sojourner, NASA's first Mars rover. Sojourner was viewed as a great success in the space community—its programming inspired by Rod Brooks' behavior-based ideas. But I liked Lee for reasons beyond his accomplishments in robotics—his quirkiness for one. Lee was a tall, burly, bearded man who spent his first career in the military. There he played fife in "The Old Guard," the prestigious Army band. Lee talked tough, rode a Harley-Davidson, and wore his heart on his sleeve.

None of us expected an easy path to our goal. Challenges involving safety, perception, navigation, and cost had tripped up previous contenders. But, with a serene confidence unperturbed by any significant experience in the cleaning industry, we were certain that we would flourish where others floundered.

Our inspiration for the robotic autoscrubber was the manual autoscrubber. We acquired a few different commercial units for the purpose of reverse engineering. Manual autoscrubbers incorporate a complex cleaning mechanism that sprays soapy water onto the floor, scrubs the floor with a large rotating brush, squeegees the resulting dirty water into a vacuum inlet for recovery, and then burnishes the floor with a second spinning brush to leave it shining. Designed to move at about walking speed, human piloted autoscrubbers run continuously for hours as they clean tens of thousands of square feet of floor. They are thus typically big, heavy, and imposing.

Big robots have always scared me. If something were to go wrong, their size, mass, and speed granted them the potential to do great harm, but—containing only the crystalized wits of their programmers—they lacked the subtleties of human perception and understanding we rely on to avoid causing such harm. Robots have many strengths—no job is too dull, dirty, or dangerous for them, and they practice eternal patience in the execution of every task. But robots' Achilles' heel is their profound lack of common sense. We must be most careful in the design and use of any robot to avoid exposing that vulnerable appendage of machine anatomy.

While describing and quantifying the benefits their products will deliver, automation manufacturers speak reverently of value. Usually, value boils down to the cost of wages customers won't have to pay when they substitute mechanization for human toil. That was the case of our Clean robot. Clean would add value to the customer's business, but because of the nature of the labor it replaced, value would accrue only slowly—at about the rate of the minimum hourly wage.

But accompanying large machines is the large risk that they might subtract value quickly and in big chunks. For example, if our Clean robot were to injure a worker or damage property by, say, driving through a display case of Rolex watches, the value accumulated during many hours of diligent service would be squandered in an instant. A single serious mishap might cost more than the robot earned during its entire tenure.

The surest defense against such catastrophes, I thought, was to make the robot small and lightweight so it physically couldn't do much damage to people or property. Probably this would require a swarm of similar robots. But practical issues made that approach difficult. For example, where is the department store manager supposed to store a robot swarm when it's not cleaning? One Clean robot could take over the cramped corner currently housing manual cleaning equipment, but that's all—all other space in the store was spoken for. So, throughout the Clean project I alternated between looking for swarm solutions where many small robots would cooperate to clean the floor (perhaps, after cleaning, climbing on each other's back somehow to stack themselves vertically) and, on the other, finding failsafe mechanisms that would ensure the harmlessness of a large robot.

PHIL MASS

Phil Mass had a superpower. He could read people. With only scant observation, Phil could unfailingly discern whether a person was a team-player or a prima donna: whether that individual could be counted on to pitch in during difficult times or would instead exploit circumstances for personal advantage. Phil's ability to peek behind the disarming façades we all erect and view a person's raw nature was a rare talent—one that's invaluable when teammates are chosen and cooperative projects are undertaken.

Nothing as glamorous as the bite of a radioactive spider had granted Phil his superpower; he earned it growing up in a chaotic household. There, of necessity, he'd had to develop an ability to instantly assess the emotions in the room. But he found refuge from the turbulence at home in scholarship. Focusing on his schoolwork and expanding his horizons gave Phil a fun, fascinating, and rewarding outlet—a welcome sanctuary that he could control. He loved learning and he excelled at it.

Ultimately, Phil left the disarray behind and arrived at college. There, he found that he enjoyed physics. But he also liked sculpture. To the confusion of his friends in each discipline, he majored in both. That came after abandoning his initial choice of a major (mechanical engineering) and switching colleges (from Michigan to Oregon). After graduating he went to physics grad school but, like me, Phil couldn't find his passion in physics and so began looking for a job. A certain restlessness seems common among roboticists.

Years earlier, Phil had happened upon an article describing Rod Brooks' robot programming ideas. Like a stubborn earworm, Rod's intriguing notions stuck in Phil's head and kept playing on repeat. Building robots behaviorally, from the ground up rather than from the top down— powered by conventional abstract reasoning—felt intuitive to Phil. The new approach seemed at once obvious and revolutionary. So Phil was thrilled when, during his job search, he discovered that Rod's company was hiring. The tantalizing possibility that he might work on behavior-based robots himself instantly shoved all competing job prospects into distant runner-up positions. Phil's zeal shone through at his interview and he got the job—his first professional employment.

iRobot hired Phil in 1998 for the purpose of writing computer code to test BoreRat, the company's oil well robot. But he never actually wrote any test code. The first day he arrived, he was put to work on more urgent matters, writing *sensor drivers*[5] for the Clean project. That suited him just fine.

CLEAN UNIVERSITY

The early days of the Clean project were filled with enthusiasm, education, and innovation. One of the things I love most about robotics is that robots are great pedagogues—they insist that their designers constantly learn new things. For every robot I've tried to build, I've needed to comprehend curious, unexpected subjects.

Clean required us to examine and understand the primary surface the robot would clean—VCT (vinyl composition tile). Our schooling included squatting on the floor and observing floor tiles under a magnifying glass— this to appreciate how previously applied layers of protective floor coatings deteriorate over time. We discovered there actually exists an instrument called a "gloss meter." This device uses light emitters and detectors to sense and report, objectively, how shiny the floor is! We also learned about the TACT circle. TACT allows one to ensure good cleaning by trading off among Temperature, Agitation, Concentration (of the cleaning chemical), and Time. Our robot might need to employ TACT to deliver the cleanest floor in the least time.

I made new friends at Clean University. Frequently I had occasion to work with Paul Sandin and discovered we were on much the same robotic wavelength. Both of us were impatient with the "gee whiz" hype that too often surrounded robots. We were keenly interested in building practical products that actually do something people want done. I also became acquainted with the many talents of Phil Mass. Along with outstanding technical chops, Phil possessed an exceptionally high emotional IQ (stereotypically only a vestigial feature of the engineering psyche).

Innovation was a constant Clean necessity. One necessity that mothered an invention was the common display hook. Stores often hang merchandise from chrome plated rods (we called them pokey bits[6]) that protrude several inches horizontally from peg boards into the aisle. Such devices attract customer attention but create a confounding hazard for robots—especially when a shiny pokey bit is bereft of hanging merchandise. The robot is virtually guaranteed to collide with anything in its path that it can't see. At the time, no conventional sensor could reliably detect a polished display hook.

Our answer was the "laser wedge wall." Jutting out from the two bottom corners on each lateral wall of our rectangular robot was a laser. The laser had a special lens that spread the beam out into a broad wedge parallel to the surface of that wall. On each wall, we placed a grid of 20 sensors designed to detect the infrared light the laser emitted. Normally, this light sailed above the sensors, undetected. But if any physical object, including a pokey bit, came within an inch of the robot's surface, light was scattered from the object onto one or more of the sensors. This told the robot that something was too close for comfort, so it could alter its course.

One early employee, Chuck Rosenberg, had left the company for graduate school by that point. But happily Chuck kept in touch, and he introduced us to a new navigation paradigm that Clean needed. The method is called *SLAM* for Simultaneous Localization And Mapping. It's an intensely mathematical technique rooted in probability theory. Using SLAM, a robot can combine many sensor readings, each containing noise and error, into a nevertheless accurate estimate of the robot's location. Curiously, SLAM emerged from the once lost and later rediscovered work of Reverend Thomas Bayes. Reverend Bayes, a native of England, lived from 1702 to 1761. He discovered a clever way to compute a challenging probability from a related, but easily computable probability. The method called

Bayesian statistics is now widely used. It fascinates me that an eighteenth-century cleric plays a critical and ongoing role in guiding twenty-first-century robots on their journeys.

Besides SLAM and the laser wedge wall, we worked on a method to distinguish between tile and carpet using high-frequency ultrasonic sensors—a robot designed to polish tile floors must assiduously avoid applying that treatment to any carpet-covered floor areas it encounters along the way.

Solving technical problems that can make a difference in the real world is something I find challenging, fun, and fulfilling. Clean contained an abundance of such opportunities. The project truly manifested all the technical treasures a devoted roboticist could hope for. But overcoming technical challenges, as satisfying as that might be, is not enough to ensure a project's success. And even as we learned and invented, metaphorical cracks and fissures were developing that no technical cleverness could repair.

TAMING THE CLEAN BEAST

A mobile robot moves when its wheels turn, rotated by the motors. The robot's high-level software decides how fast each motor ought to rotate. But how quickly it actually does spin depends on two things: 1) the amount of power sent to the motor, and 2) the *load*. The phenomenon of load will be familiar to anyone who has ridden a bicycle uphill or across sand, it corresponds to the effort you have to expend to move at a chosen speed as circumstances change. When a motor spinning at a constant speed experiences an increase in load, it slows down, just as you might upon encountering a hill.

Driving about, a robot meets different conditions that change the load on its motors. So, if we want to keep the robot's velocity steady, we must constantly adjust the power flowing to its motors. A bit of essential software known as *motor control code* has this job. Motor control code compares the speed we want a motor to spin with how fast it actually is spinning and tries to achieve a match. If the motor is turning too slowly, the code sends it more power; too fast, it sends less. (That sounds simple but there's an entire branch of engineering called *control theory* that studies how best to do this.)

We've all experienced computer crashes and glitches. Always annoying, they sometimes manifest as a popup window that displays an unwelcome message, other times the screen might scintillate with odd colors and patterns. But what happens when the computer controlling a robot experiences a similar snafu?

Stuffed with massive batteries, powerful motors, and heavy-duty cleaning hardware, the prototype Clean robot tipped the scales at about 600 pounds when Phil Mass and a couple of other engineers rolled it into iRobot's basement test area. Their purpose was to try out the new motor control code Phil had been working on.

With his laptop mounted on the rear of the Clean robot, Phil connected a cable and loaded his code into the machine. Then he typed a command that would make the robot drive forward. To play it safe, he specified a sedate velocity.

Unnoticed by Phil was a small error in his algorithm—a plus where he'd meant to type a minus. The effect on the motor control code was that sometimes, when it should have sent less power to the motors, it sent more.

The instant Phil hit the return key bedlam erupted. The heavy robot began making violent and unpredictable motions. Repeatedly, it lurched forward, spun in a tight circle, then lurched again.

One wall of the robot test area was made of glass—frosted for privacy on the side where it ran along a hallway. Across that hall another glass wall, this one clear, formed part of the perimeter of a gym. Unsuspecting fitness enthusiasts absorbed in their routines toiled away in this space, blissfully unaware of 600 pounds of whirling doom separated from them only by fragile glass and thin air.

Desperate to avoid becoming the lead story on the nightly news, Phil typed frantically at his computer, working to correct what he now knew to be faulty code. This was tough because the uncooperative robot kept swerving away from him. So, again and again, Phil was forced to type a couple of characters on his laptop, chase the robot down, and then type a couple more.

After several minutes (!) of this madness, Phil was finally able to repair his code and calm the raging robot. Remarkably, amid the chaos the robot had remained clear of the wall, injured no one, and failed even to damage itself.

You might be wondering, "Why didn't Clean's electrical engineers provision the robot with a BIG RED EMERGENCY STOP BUTTON?" This simple feature could have prevented the entire incident by aligning the robot with a safety standard replicated on nearly every piece of industrial equipment on the planet. The answer is, they did. The robot's emergency stop button was well within his reach and Phil could have pressed it at any time.

It just never occurred to him to do so. Nor did it occur to any of the other engineers in the room with him.

I understand Phil's button blindness—I'm a fellow sufferer. It seems we robot developers inadvertently learn to ignore emergency stop buttons. In just the same way that you can lose your work when you reboot a misbehaving computer, pressing the emergency stop button on an unhinged robot can erase crucial information needed to diagnose the very problem that created the emergency. Developers urgently want to avoid this. Thus we internalize the notion that the emergency stop button is a feature meant for someone else, taboo to us developers.

Note that in this anecdote, an error of only a single character in one piece of code transformed what was intended to be a docile servant into a raging beast. Phil was confronted with his error the moment he tried to run the robot. But what if the mistake had been hidden more deeply, revealing itself only under rare circumstances during customer use? Did I mention that big robots scare me?

THE SPRING OF OUR DISCONTENT

As development proceeded, the Clean robot became ever larger and more complex. Achieving the technical solution for the price SCJ wanted proved more difficult than we had supposed. (Sadly, my joke from two years earlier was prescient. We really should have asked for more money.) Over time, more and more people were assigned or loaned to Clean from other projects. Lee Sword bragged that at one point or another essentially every iRobot employee had worked on Clean. But rather than accelerate progress toward a common goal, the extra hands seemed to pull in conflicting directions. Squabbles emerged; joy faded.

Dropping in on a meeting of the full Clean team just a few weeks before a major demonstration in Racine we find an all-too-frequent dynamic playing out. The lead mechanical engineer is complaining, two electrical engineers are bickering, another mechanical engineer is sobbing, and our SLAM expert has just stormed out. Presiding over the kerfuffle of loud voices and ruffled feelings, Lee appeals for calm and patience … to no lasting effect.

As quarrels and delays accumulated Lee sought solutions. Informed by his military background, Lee added more and more levels of hierarchy to the team. But the girders and beams of new management structure failed to corral his creative but too-often cocksure engineers.

I found our situation uncomfortable and confusing. We were privileged to work on a great project, and we clearly had the talent to succeed.

The popularity of robots in the tech community worked strongly in iRobot's favor, enabling us to hire only the most accomplished candidates. Several times I participated in interviews where, unbidden, potential hires brought along robots they had designed and built themselves. (These proactive folks always got a job offer.) Other times candidates accepted an available position for which they were overqualified just to get in the door. We used to refer to such extreme enthusiasts as having been bitten by the "bug." The only known treatment for the resulting affliction was working on robots. (Paul, I, and many others at iRobot were confirmed victims.)

To a person, Clean team members were experienced and highly capable, our leader sincere and enthusiastic. Yet Clean was not a happy team. Frequently Lee confided in me, relating details of the dysfunction we've descended into. "What can I do to keep the peace?" he pleaded. But I was no help; "I don't know," was all I could say.

WINSTON TAO

Young Winston sat transfixed. Plodding step by step across his flickering TV screen, he watched the little robot make its way. The size of a lunch box, all exposed metal and gears, the simple robot carried no payload. Except dreams.

Every week precocious six-year-old Winston Tao tuned in to his favorite grownup TV show, *The 21st Century*, for a glimpse into the future, his future. The episode, broadcast March 12, 1967, would prove life changing. Legendary newsman Walter Cronkite, the host, promised, "Robots are coming. Not to rule the world but to help out around the house." Professor Thring of Queen Mary College, London, robot-inventor and guest, explained that one day robots like his would do all routine housework. And they needn't look like people to accomplish this. Domestic robots might have three or four hands or eyes in their feet!

The prospect thrilled Winston and ignited a spark that would not fade away. "How neat would it be to build something like that!" Winston thought.[7]

Years later, in college, Winston majored in physics. He went on to study geophysics in graduate school. There he focused on the mechanisms behind plate tectonics, the powerful but ponderous geological forces that thrust up mountains and spread the sea floor. But early in graduate school Winston became impatient. Although the field was fascinating, he feared that the progress of his career would likely keep pace with the motion of the continents. "Science is too slow for me," he concluded. But he nevertheless stuck with the program and Harvard awarded him his PhD a few years later.

Seeking something less glacial, Winston tried government. His credentials and persistence earned him a prestigious spot on the personal staff of Representative George Brown of California, Chairman of the House Committee on Science, Space, and Technology. During Winston's congressional tenure, iRobot contacted another staffer in Congressman Brown's office to arrange a visit (as mentioned earlier). Through the most tenuous of circumstances Winston encountered Colin, Helen, Jill, and me when we visited Washington on that day in 1994. (Winston intervened to make sure Representative Brown saw our demo; normally, we would not have gained an audience with an actual member of Congress.)

Winston learned much in his new position and was greatly impressed by the intelligence and thoughtfulness of most of the people he worked with. But there turned out to be a fly in this ointment as well—government had the opposite problem from science. On Capitol Hill, the weekly onslaught of bills and debate was rapid and relentless, often leaving Winston wishing he had more time to study the issues in depth. "The pace of government is too quick for me," Winston thought.

Looking for a Goldilocks solution, Winston started business school in 1997. His institution, Wharton, like many others, encouraged students to seek internships with businesses during their studies. Winston and Helen had kept in touch ever since iRobot's visit with Representative Brown. Because of that and his childhood fascination with the robot on *The 21st Century*, Winston interned at iRobot. The experience was positive and, returning to business school, Winston focused on developing knowledge and skills that would be helpful to a business like iRobot's. When he graduated, Winston turned down a much more lucrative, competing offer from a prestigious firm to join iRobot as their first vice president—he still wanted to build robots. The company had about 40 employees when he joined.

CLEAN OUT

On his first day at iRobot, fresh out of business school, Winston Tao was assigned to replace Lee Sword as director of the Clean project. This was a surprise to Winston (not to mention Lee) who had expected to spend his time analyzing business strategy rather than leading a product team. Time was short as we raced to prepare for the critical demo in Racine.

A few days before the big event a couple of iRobot employees packed the big Clean robot into a van and began the 1000-mile drive to Wisconsin.

Winston, Paul, Phil, several others, and I flew in to meet the robot and set up for the demo.

Key systems were unfinished. Our SLAM navigation scheme didn't function. This meant the robot couldn't find its way around large spaces. Whenever Clean turned a corner, dirty water spilled off the edge of the squeegee that was supposed to guide it into the vacuum inlet, leaving an embarrassing puddle behind on the floor. The laser wedge wall obstacle avoidance sensor of which I was so proud could be fooled by unintended, distant reflections. And critical mechanical components of the robot kept breaking. We were not ready for prime time.

But to the great surprise of both Winston and his counterpart at SCJ (who'd been tracking our troubles) the demo ended in applause rather than disaster. By carefully engineering how we conducted the demonstration we were able to show the robot in its best possible light—its many warts invisible to the casual observer. Still, Clean was far from mastering the skills it must have to do its job.

SC Johnson contracted iRobot to build a robotic autoscrubber because they saw automation as key to a new business venture. But shortly after the Racine demo we received word of a change. The official explanation was that SCJ had reevaluated the business they were pursuing and decided that even with the robot the venture would not be viable. Without the business, there was need for neither the Clean robot nor the Clean team.

Ambivalence was my salient sentiment as our project expired. I was disappointed that we wouldn't be the first to market an industry-defining robotic autoscrubber but relieved that the uncomfortable meetings and perplexing conflicts were over. A small part of the Clean team's efforts lived on. Commercial cleaning equipment manufacturer Tennant took over the Clean patent.[8] Tennant packaged the non-robotic elements of Clean into a manual three-in-one autoscrubber called NexGen. NexGen was sold for several years.

Lee Sword rebounded from his abrupt dismissal and went on to lead other projects at iRobot. His later efforts, mostly involving robots for the military, reached much more satisfying conclusions.

Regrouping after Clean's demise, Paul and I conducted an informal postmortem. How could such an interesting project, brimming with promise and passion have gone so wrong? Was the Clean team just too big? Were incompatible personalities the culprit? Were we doomed by immature technology impossible to perfect in the time available? We couldn't dig down all the way to the root of Clean's ruin, but we resolved in future to avoid every pitfall we knew of or suspected.

Next time, we decided, the team should be small, the management structure flat, and we'd favor compatible personalities over individual stars. We'd also try to avoid robots that needed too much new technology all at once.

Paul and I weren't alone in our contemplative reaction to Clean's end. Separately, Phil too was pondering team dynamics. His readings on the subject suggested that Clean's experience—where adding people retards progress—was all too common in the corporate world. Next time, he would look to join a small, cohesive team with big goals. One whose lean roster held no place for troublesome recruits.

Neither Paul, Phil, nor I would have to wait long to put our ideas into practice.

NOTES

1 The antagonist in a horror film franchise, Chucky is a malevolent child's doll.
2 Some reasons for worker unhappiness are listed here: https://cybercleansystems.com/robottsunami.html.
3 RoboKent was developed by Transitions Research Corporation with funding from Electrolux. It was appropriate for cleaning long hallways but not open areas or spaces with complex geometries.
4 An actuator is any powered, controllable mechanism used to make something move. Electric motors often serve as one type of actuator. A hydraulic actuator functionally resembles a syringe. Only instead of pulling or pushing the plunger to make the fluid flow in or out, the mechanism adds or removes fluid to make the plunger move.
5 A sensor driver is a small program that controls a sensor. This bit of code confines the idiosyncrasies and peculiar requirements of each sensor to its driver code. That simplifies the sensor's interface to the robot's main program.
6 Curiously, although soon-to-be-introduced Roomba developer Chris Casey was not a member of the Clean team, he coined the "pokey bits" term that we all used.
7 At the time, Winston lived with his parents (both refugees from the communist takeover of China) and his sister in a tiny apartment in Queens. Winston's parents encouraged their children to learn and explore.
8 Patent US7240396B2. The machine swept, scrubbed, and burnished all in one pass.

CHAPTER 7

Coverage

C ONSUMERS EXPECT ROBOTS TO behave like people, only more precisely and efficiently. They want to see a floor-cleaning robot travel in straight lines, execute sharp right-angle turns, and clean effectively—it especially shouldn't re-clean spots it's already cleaned. But around the turn of the millennium those were lofty challenges. Why? Because they required the robot to know where it was, a problem called *localization*.

Knowing where you are within a room seems obvious and trivial. Just keep your eyes open! Remembering where you've been is equally easy for us; our innate abilities let us do both, making efficient vacuuming a breeze. Today, thanks to the aforementioned SLAM technique and the ever-falling price of sophisticated sensors and processors, robots *can* localize. They mostly succeed in recognizing where they are[1] and remembering where they've been. But a couple of decades ago such feats were on technology's cutting edge, prohibitively expensive for a consumer robot. Only eyes-closed forms of navigation were obtainable.

If vision isn't available for localization, there's another trick in our magician's bag that might help. Called *dead reckoning*, it was the favorite method of early, would-be robot vacuum inventors. As a robot moves, its wheels turn. What if we count the rotations? An encoder, a type of sensor enables this. Using the known diameter of the robot's wheels, the distance that separates them, and how far each wheel has turned we can—by applying a bit of geometry—calculate the distance and direction the robot has moved from its starting position. In theory, this enables the robot to always know exactly where it is relative to where it started. But the real world shows robots no affection. In practice this doesn't work.

DOI: 10.1201/9781003540489-7

Why is dead reckoning a mirage? You can intuit the answer yourself with a fun experiment. Start outdoors in a safe place far from roadways, stairs, or other hazards. Mark your starting point in some way, say using a leaf, stick, or small rock. With your eyes closed *carefully* take 15 steps forward, turn 180 degrees, and walk 15 steps back. Open your eyes. Did you return to your exact starting position? If you managed to get pretty close, try increasing the challenge by walking in a large square 15 steps on a side. You'll find that the more distance you cover with your eyes closed, the bigger the error you'll observe when you open them at the end. This can help you empathize with the robot that has no eyes to open.

On a hard, flat surface over a short distance dead reckoning can work pretty well. But as the length of the robot's excursion increases the difference between where it thinks it is and where it truly is just gets bigger and bigger. Wheels slip, errors add up, and pretty soon the robot is hopelessly lost.

So, is our cleaning task equally hopeless? Not at all! The robot just can't accomplish it the same way a person would. Remember that the problem we actually want to solve is not localization, but coverage. To clean the floor the robot must cover it; that is, it must visit every point on the floor that it can physically reach. It turns out the robot can guarantee full coverage without ever knowing where it is. How? Embrace chance.

Suppose we implement this algorithm: The robot drives forward in a straight line (as nearly as it can). Then it bumps into an obstacle. The robot picks a random angle between –180 degrees and +180 degrees. It turns that many degrees (as nearly as it can). Then drives forward. Repeat.

That simple algorithm is enough to guarantee complete coverage. The robot never knows where it is, yet it cleans the entire room, eventually. This method is not efficient, but it has two properties that make it attractive to a floor-cleaning robot. It's certain to work and it's cheap.

"But," you may object, "if the robot is merely bouncing randomly how can we be sure it will clean everywhere?" In the same way we can be sure that all the air molecules in a room won't end up just bouncing back and forth on the left side, never finding their way to the right where you sit gasping. Our shining knight, Sir Statistics, has our back. It's mathematically inevitable that if the robot moves randomly, over time it will visit every spot that is physically reachable. Every casino in the world profits by betting on the certainty of chance and the robot can do the same.

As always, there is some subtle complexity. The catches are "over time" and "randomly."

What if the robot recharges its batteries at a fixed charging station? In this case the robot always starts its cleaning routine from the same spot. That's a problem. Imagine that the arrangement of furniture in a room essentially creates a virtual labyrinth, one that requires the robot to bounce in just the right direction at a number of different spots to reach the far side. Were the robot to run indefinitely this would always eventually happen. But, if every time the batteries start to fade, we make the robot start over from the beginning, it may have trouble reaching the area on the far side of the maze.

A well-known trick can be used to solve most mazes. As you enter, put your right hand on the right wall (left on left also works). Then sally forth while never removing your hand from the wall. Eventually, you will exit the maze. The robot can use a similar stratagem. When it encounters a wall, sometimes it bounces away and sometimes it follows the wall for a short distance. This has been found to be an effective way for a robot to cover cluttered (maze-like) spaces.

Unfortunately, just because we command the robot to move randomly doesn't mean it will. Loose floor covering, odd geometry, and other features can conspire to make the robot repeat the same action even though it's trying to perform different ones. This non-randomness can lead to coverage failures where parts of the space are starved for attention while other parts are over-covered.

We coined the term *systematic neglect* to describe such situations—where coverage is incomplete because the robot *doesn't* behave randomly. No amount of math can prevent systematic neglect. All one can do is to test extensively to make sure it isn't happening and then change something if it is.

NOTE

1 This is usually true indoors. To localize visually, a robot must be able to see lots of unique visual features that largely stay put. But even accompanied by the spirit of Reverend Bayes, a robot may become lost if few reliable features are visible (e.g., in the middle of a dune-less desert) or if the same features endlessly repeat (e.g., a sea of office cubicles all the same).

CHAPTER 8

Forging a New Machine's Soul

I T WAS A COOL Monday evening in May 1999. I was at home still pondering the unsatisfying end of the Clean project. After finishing dinner with the family, I checked my email and found a message from Paul Sandin. Writing in his excited way, Paul described an epiphany he'd just experienced while vacuuming. Thinking about how we might blaze a way forward from Clean's disappointing conclusion, it suddenly occurred to Paul to invert the scale. Many of Clean's problems seemed rooted in the robot's great size and complexity, so what if we went the other way? Paul proposed a new robot that would be small, lightweight, and simple in the extreme. It would be cylindrical, incorporate an instrumented bumper to detect collisions, and use differential steering (the same scheme that Rug Warrior had used). Relying on a carpet sweeper mechanism rather than a vacuum would make it feasible!

I smiled broadly while reading Paul's message. It delighted me that we'd independently reached the same conclusions about the best way to build a home robot floor cleaner. But I wasn't instantly supportive of Paul's proposal. The first thought to mind was more like, "Been there, done that." Despite having worked closely with Paul for over two years on Clean, I'd somehow never shared my stories about Rug Warrior, RoboBroom, or the hoped-for GoldStar/Bissell-bot. I corrected that omission the next day at work. It's embarrassing now, but the tone I projected at the time conveyed an attitude of don't-get-excited-I-already-tried-that.

    DOI: 10.1201/9781003540489-8

"Maybe it's time to try again," Paul pushed back.

Hmm ... The more I considered it the more I thought Paul might be right. The company was in a very different position from the days when Bissell and GoldStar rejected us. iRobot's survival was no longer threatened on a month-to-month basis, and we'd learned useful lessons about manufacturing and minimizing cost. We had also accumulated anecdotal evidence that people might want the robot we hoped to build. One thing hadn't changed—robot floor cleaners still resided exclusively in Futureville.

The market data that convinced us robot vacuums might be popular arrived informally and unsolicited. When I met someone new at a party or other social gathering, the interaction usually followed this script:

Party guest: "... and what do you do?"
Me: "I design robots, I'm a roboticist."
Party guest: "That's so cool! Can you make me a robot that will clean my floor? Ha, ha."

Occasionally, instead of "clean my floor," my new acquaintance might mention their house, kitchen, or bathroom. But the request for some sort of cleaning robot was uncannily common and consistent. Other roboticists I knew reported similar encounters. Many people, it seemed, didn't enjoy repetitive cleaning chores, and believed that a robot could help them. I thought so, too!

Paul and I both had long expected great things from mobile robots and were frustrated that their promise remained so largely unfulfilled. Over the years, we'd seen many stories in the press that tantalized but never delivered.[1] Reports often followed this breathless format: "Company *X* just demonstrated a prototype of their revolutionary robot vacuum cleaner! *X* will begin selling their exciting new robot next year." But somehow, next year never arrived. Eventually I came to suspect that the fault lay not in earth's orbit of our star, but in ourselves.

Technologists love technology—me included. But for too many roboticists, I thought, that love elevated technology to an end rather than a means. When this happened their robots became vehicles for showcasing high technology rather than products built to satisfy a human need. I found that often the most embarrassing question I could ask a fellow roboticist was, "What does your robot do?" My peer might discourse at length about processors, sensors, software and so on, but then stumble over exactly what compelling purpose that fusion of components was intended to serve.

Paul and I spent hours trying to understand robots' disappointing progress. We regarded autonomous vacuum cleaners as the lowest hanging fruit on the tree of robot possibilities. We lamented, "All they have to do is move around and not get stuck!" It just didn't look that hard. So why had no robot vacuum ever achieved commercial success?

Cost seemed to play an important role. Stories about just-around-the-corner robot vacuum cleaners always emphasized their cutting-edge technology. Paul pointed out that fancy technology would inevitably make such robots expensive. I agreed, opining that no one would pay ten times the price of a manual vacuum for a robot cleaner just because it was a robot.

We debated such matters at every opportunity. Finally, after much back and forth, we declared victory. We convinced ourselves that we grasped the subtle errors every other would-be robot vacuum inventor had been making over the past few decades. Furthermore, our new understanding pointed to an approach we were certain would succeed. (I'd attribute such hubris to the arrogance of youth, but I was 46 at the time.)

In a nutshell we concluded that we must treat the robot as a product not "the future." And that the robot's appeal to customers must be based on what it does and what it costs, not what it is. These thoughts, unconventional at the time, matured into three principles.

First and foremost was affordability. If we wanted customers to see our robot as a reasonable floor-cleaning choice, its price had to be competitive with other floor-cleaning solutions. In those days, many manual vacuums were available for $50 to $150. That established an unyielding anchor. We figured customers might pay a little more for the convenience the robot provided, but not a lot more. A second advantage of a modest price was that it lowed the stakes. People should be more willing to accept the risk that a first-of-its-kind product wouldn't work perfectly if the experiment didn't cost them too much.

Second, ease-of-use was mandatory. We were selling convenience, so the robot must truly be convenient. Ideally, a user would just press the button, and the robot would clean the floor; asking users to exchange a floor-cleaning task for a robot-wrangling job would negate the appeal. The machine should require no setup, no supervision, and certainly no programming. The only assistance we'd allow the robot to ask of its user was to dump the collected dirt.

Third, minimize technology. This stipulation contrasted with most other efforts in robotics. But as Paul suggested in his initial message, our design needed to be simple to the point of being dumb. We were willing—eager even—to use established technology in innovative ways, but the robot's basic components should be common off-the-shelf parts, not the cutting-edge constructs described in research papers. *Low* technology, we believed, would help make the robot reliable and let us meet our cost imperative.

Taken together, these principles were actually quite challenging, maybe even self-contradictory. For example, designing a device that was simple and intuitive to use might require a great deal of hidden complexity, i.e., lots of technology. Had we been more circumspect we might have wondered whether a solution that satisfied all three principles simultaneously was even possible. But we brooked no such doubts: Full speed ahead!

DUSTPUPPY

Like the guest of honor at a chickenpox party, Paul and I needed to infect others with our mind virus. But we feared iRobot was becoming a company of party-poopers—increasingly immune to daring ideas. The egalitarian commune of dreamers we'd once been might now be stifled by corporate hierarchy. Gaining favor with the new order of CEO, CFO, and other C-sovereigns would require a speaker fluent in their enigmatic tongue, we thought.

In VP Winston Tao we saw our salvation. Thrust unexpectedly into the leadership of the Clean project, Winston stepped up. Displaying a cool head, intelligence, and patience he calmed a storm of overwrought passions and guided the team to a better-than-expected outcome. We figured with savvy Winston as our champion we could not fail.

Successful suitors sometimes serenade their subject. So we poured our hearts into a comprehensive document for Winston. We aimed to anticipate and then answer every pertinent question. What does the robot do? How does it work? Who will use it? How will they use it? How much will it cost? And importantly, what will we call it?

New technology always comes to bat down two strikes. New tech is raw, unrefined, and often confusing to users. Old technology is optimized, polished, and familiar. Even when your new invention has the potential to outpace a venerable standard, like a smartphone competing with a landline, rough edges and misunderstandings can make the new tech look like a dud. Manual vacuum cleaners, our competitors in this contest, had 100 years of incremental improvements under their belts (so to speak). We worried that our robot's temporary imperfections would make it hard to love. How could we manage expectations?

Eventually it struck me, our robot was a lot like a puppy. Filled with enthusiasm and desperate to please, a puppy always tries its best. But sometimes it messes up. If we could get users to think of their robot as well-meaning but immature, they might be more forgiving. The name DustPuppy seemed to evoke the right sentiment.

Focused on thoroughness, we progressed slowly and cautiously. It took nearly six months but finally our document titled "DustPuppy, a Near-Term, Breakthrough Product with High Earnings Potential" was ready. It answered all the questions we thought crucial. What does the robot do? It cleans home floors. How does it work? Using a carpet sweeper mechanism it will bounce randomly until the floor is clean. Who will use it? It will be helpful to pretty much everyone, especially suburban homeowners, harried parents of young children, frugal apartment dwellers, seniors who have trouble wielding a regular vacuum, and (gifted by their parents) kids in college. How much will it cost? Our price will be just under $100.[2]

We described usage scenarios, the way previous robotic attempts had gone wrong, and options for distribution and promotion. We touched on market size, development schedule, and follow-on products. And we concluded that iRobot was very well positioned to undertake such a project, that it would greatly benefit the company, and that we should begin development being immediately. Go big or go home, as they say.

Paul and I were probably too optimistic about how quickly DustPuppy could be developed, how well-positioned our company was, and how likely we were to succeed. But we were eager and excited and thoroughly convinced we'd found the Holy Grail of mobile robotics. We would build a new robot, and it would sell hundreds, maybe even thousands of units! Our proposal alluded to millions of dollars in sales iRobot might reap from DustPuppy, but it was tough for us to picture that actually happening. The most successful mobile robot of the era, a hobbyist robot from HeathKit called the Hero 1, surpassed 10,000 units[3] only after several years on the market. The foreknowledge that DustPuppy would outpace Hero a thousandfold would have left us stupefied.

On a mild day in early December, toting our proposal along with other supporting documents, Paul and I walked down the hall to Winston's office and knocked on the open door. Winston looked up from his desk and I asked, "Do you have a minute? We've got something that might interest you."

Always gracious, Winston said he did, so we entered and began our pitch. I told a long story. It described the robot we'd wanted to build for

Bissell or GoldStar and the much simpler robot we had originally intended for Clean before it grew into a behemoth. Then I went through our DustPuppy proposal pointing out how useful the robot would be to the company and why, because of the many lessons we'd learned, we were now in a great position to succeed.

Only much later did I discover that our visit was far from unique. Engineers frequently dropped in on Winston to harangue him with their pet robot ideas. He always listened patiently. But invariably Winston found these unsolicited concepts to be either poorly conceived or devoid of market value or both. Afterwards he would quietly let the dubious ideas die of their own ill health.

But, as we finished our presentation, Winston seemed intrigued. None of his questions stumped us, and his comments suggested that wheels were turning. Sensing Winston's lack of immediate opposition, I pounced. "Give us $2000 for materials and in three days we'll have a working prototype on your desk!"

Winston knew better. He had learned the foibles of engineers and understood the need to apply a multiplier, a sort of optimism correction factor, to any number a member of our class might propose. He assigned an even larger correction when Paul and I asserted the number. But despite our rose-colored projections, we had succeeded. Our evidence and erudition had convinced Winston that our concept was realistic and worthwhile. Wanting to give our robot a fair shot unencumbered by our reckless enthusiasm he said, "Joe, it's going to take you two weeks and $10,000. Just make sure it's good enough to show Colin."

TWO WEEKS AND $10,000

So, the dog caught the car. I'd been imagining a longer process. First, we'd build something quick and dirty that only Winston would see. If he liked that robot (and I was sure he would) we'd then spend additional time refining our concept before exposing it to the judgment of top management. But Winston short-circuited that comfortable notion and raised the stakes considerably. In just one meeting DustPuppy transformed from Paul and Joe's private castle in the sky into a fully visible, officially sanctioned company project with a budget and a schedule. We sipped from a cocktail of giddiness laced with terror.

In two weeks' time we must deliver a fully baked prototype. While fantasizing kudos and high fives I'm painfully mindful that the last time I demonstrated a little floor-cleaning robot to a company's top brass things did not turn out at all well. Our new robot needed to be considerably better than "good enough to show Colin." This time even skeptics must come away convinced that we're on to something game changing.

Paul and I had to draw on every lesson we had learned about designing robots, collaborating, limiting scope, and so on. The coming fortnight was suddenly fully booked. We had to get this right.

Our project was called DustPuppy, but we gave the first prototype a name of its own. Inspired by a character from Disney's *Lady and the Tramp*, we settled on Scamp. I used to enjoy the adventures of that irrepressible little gray dog in the comic pages of my local newspaper when I was a kid.

The plan was to make Scamp work well, to develop it quickly, and to demonstrate that the production robot will be low in cost—we needed the trifecta of good, fast, and cheap. Inauspiciously, an old engineering proverb warns that of that trio—important to almost every project—you are allowed to pick any two. Harsh reality always forced you to trade away one attribute to obtain the remaining pair. But to prove our approach, we really needed to show all three. How could we finesse this?

Years of building robots had taught me some valuable strategies. An important one was to cheat whenever possible. (In a principled way, of course!) Our deadline was fixed but—begging the indulgence of our families—we could steal a little more time by igniting both ends of the candle. We'd be working nights and weekends for the duration.

Do one thing well is an imperative I try always to heed. Audiences cheer when a robot performs a single task skillfully. But gambling for greater glory by showcasing multiple talents can backfire. Applause turns to crickets when the performance of any feat falls short of virtuosity. In robotics it's always wise to favor quality over quantity, competent depth over mediocre breadth.

We'll limit Scamp's scope to improve our odds. Of the many things we want DustPuppy eventually to do, Scamp will present only a carefully chosen few. Focusing our efforts grants us time to perfect key abilities. Doing one or a few tasks well would also earn our audience's trust—seeing that everything we demonstrated worked well, they would give us the benefit of the doubt that we can perfect any abilities we didn't show.

Achieving rock-bottom production costs, as My Real Baby taught us, involves using cheap, *in*efficient motors coupled to custom-designed, injection-molded gearboxes. But given Scamp's schedule, that formula was beyond our grasp. The time to design a mold and the expense of building it would blow up our budgets for both days and dollars. Instead we would use a pair of high-efficiency *gearmotors* (preassembled motor/gearbox combinations) from the stockroom. Our justification: Later, once DustPuppy had proven itself worthy, a full engineering team could implement the lower-cost method.

Our sprint began with high hopes, a keen focus, and good wishes from our colleagues and community. One family member even made a tangible contribution. "My dad wants to help," Paul told me. "He'll make a mold for the shell."

Scamp needed a lightweight, round, flexible shell. We planned to fabricate one using a technique called *vacuum forming*. This involved stretching a thin, flat piece of plastic over a solid shape (the mold), heating the plastic until it became pliable, and then exposing the mold-side of the plastic to vacuum. This forced the plastic to collapse onto the mold, permanently assuming the desired shape. (A vacuum forming mold is vastly simpler and maybe 1000 times less costly than its injection molding counterpart.)

In short order, Paul's father delivered a beautifully turned and polished maple disk made to our specifications. A roughhewn chunk of wood would have sufficed, but Paul's father, a true artisan, was incapable of producing something so crude. Unfortunately, we neglected to describe one important feature of the mold—we needed a big indentation in the top to create a hollow across which we'd mount the handle. Sadly, we had no choice but to cut into the mold. Paul used a machine tool to gouge out a cavity in the top of the maple disk, destroying the beauty of the elder Sandin's work.

Paul sculpted Scamp's chassis from pieces of sheet metal that he cut, bent into the proper shapes, and then fastened together. We sacrificed a Bissell carpet sweeper on the altar of robotic advancement to obtain the brush and other mechanisms Scamp needed to sweep up dirt. Paul mounted the plastic shell on the chassis, instrumenting it with small electrical sensors we called "bump switches" so Scamp could tell when it ran into something. He manufactured two drive wheels by machining metal disks. Around the disks he stretched large rubber O-rings to serve as tires. The wheels attached to the output shafts of our gearmotors, the bodies of the motors fastened to the chassis. At the front of the chassis, Paul mounted a small, spherical roller made of Teflon. This gave Scamp its balance. The smooth roller also let Scamp glide and pivot easily as it maneuvered. The transplanted brush from the Bissell was situated between the two drive wheels. Paul powered it with a dedicated motor connected to the brush via a tough rubber belt.

Like the Scarecrow in *The Wizard of Oz*, Scamp needed a brain. But a twister-powered trek to the Emerald City proved unnecessary as I could mail-order a suitable microprocessor-based computer board from a company in Texas.[4] The board provided several input and output lines. These electrical links enabled the program running in the microprocessor to measure and effect the physical world. We wired input lines to the robot's bump switches and other sensors. Output lines gave the microprocessor control of the motors. Just like the Clean robot, Scamp needed motor control code. But there was no possibility that this robot would reprise Phil Mass's harrowing escapade. If (when) I made a coding error that deranged Scamp's mind, I would just pick up the robot like a misbehaving two-year-old.

We endowed Scamp with multiple personalities. A labeled, three-position switch allowed us to choose which one the robot would channel. The first was "Spot"; it was intended for cleaning up spills or isolated patches of dirt. Spot made the robot spiral outward from its starting point until it had cleaned a circle about three feet in diameter. Dead reckoning, as described earlier, enabled this behavior. However, errors inherent in the method limited the size of the spot Scamp could reliably clean. If we specified a too-big diameter, Scamp's path would begin to resemble the walk of a drunken sailor rather than a diligent cleaner.

"Edge" was Scamp's second persona. It made Scamp obsess about cleaning the perimeters of walls and objects. Edge's simple algorithm worked this way. The robot first arced forward and toward the right—toward a wall. When Scamp touched the wall the bump sensors activated, and the robot stopped arcing. It then spun to the left about its center for a short time. Afterwards the robot arced forward and toward the right again. If Scamp was adjacent to a wall when Edge was initiated, these steps, repeated over and over, caused the robot to hug the wall as it forged ahead.

The final personality, "Room," was the most complex and most useful. Starting Scamp with the selection switch in the Room position made the robot first spiral outward as with Spot. But when the spiral reached three feet in diameter the robot changed tactics and took off in a straight line. Eventually, Scamp would bump into something. When this happened it spun in place a random number of degrees and then resumed driving forward. Sometimes, instead of spinning, the robot followed the wall as Edge would. But after a short distance it turned away from the wall and resumed driving straight. Eventually, after behaving in this way for long enough, Scamp would clean the entire room.

DEMO

We planned, we built, we programmed, we tested, and we iterated our design. Two weeks hurtled past in a blur. But Winston's estimate was accurate. When demo day arrived, we were ready.

Colin, Helen, Winston, Paul, I, and a few others gathered in a space that Paul and I had outfitted for robot testing. A dense, low-pile carpet covered the floor—the same surface on which we had done all our testing. Wooden boards standing on their narrow sides outlined a 100 square foot test area. We had placed chairs, a small table, and other obstacles in our "typical" room. Scattered over the floor were carefully selected particles of synthetic dirt.

Speaking to our audience with nervous haste, I described our vision, our progress, and Scamp's personalities (behaviors). We demonstrated each in turn. Spot. Scamp spiraled outward and stopped at the programmed diameter, cleaning up the artificial dirt along the way. Edge. Scamp circumnavigated the test area; I stopped it when it returned to its starting point. Room. Scamp spiraled, drove straight, bounced away from obstacles, and followed walls.

I let the Room behavior continue running while Paul and I promoted Project DustPuppy and answered questions. Captivated, all present gazed intently at the robot as it went about its programmed task. Humming and brushing sounds provided self-accompaniment to Scamp's improvised dance.

The robot occasionally missed a speck of dirt the first time it drove over but by the end of the demo the floor was spotless, and Scamp had avoided getting stuck or even acting oddly. Paul and I devoted no time to explaining what Scamp was supposed to do because it had simply done everything we asked of it. This demo had gone as well as any in my long history of robot presentations.

Paul and I were dismissed.

After an eternity of maybe 30 minutes we met up with Winston. "DustPuppy's a go!" he told us. No additional research, documents, tests, sweat, or blood would need to be extracted. Winston reported that as soon as Paul and I left the room, Colin thought briefly, and then decided on the spot to authorize the project. Cool!

A moment of celebration was definitely in order. But realistically, this was not a time for self-congratulations. Having our project accepted meant that we had advanced to the same starting point reached by dozens before us. Ahead the path became more treacherous, and the stakes grew higher. Bivouacking at Everest base camp carried no assurance that we would ever stand on the summit.

But at least the curse of RoboBroom was broken. Searching for a new job was not in my immediate future. All we needed to do now was to skirt the boundless robot burial ground that had become the final resting place of all past attempts.

NOTES

1 We were aware that robot vacuums had a poor track record. But research for this work uncovered a much larger number of failures than either of us knew then.
2 We missed on that one. Our motivating idea was that the robot's price should be competitive with manual vacuum cleaners, thus we aimed at $100. We tried mightily, but ultimately, $199.95 was as close as we could get to that target.
3 https://en.wikipedia.org/wiki/HERO_(robot).
4 Axiom Manufacturing.

CHAPTER 9

Tango?

A S THE WORLD RANG in the new millennium, we began serious development of DustPuppy. But what exactly did serious development mean? Paul and I had already constructed a proof-of-concept robot, Scamp, that cleaned a floor. We demonstrated our device to management, and everyone was satisfied with what they saw. Why not just write up what we built, email that design to a factory, and request a million copies?

Our proof-of-concept prototype was the trailer, not the movie. The prototype presented only a rough sketch of the product to be. First, our design was incomplete. Scamp lacked key elements like conveniently rechargeable batteries, easily emptied dust cup, and the ability to avoid falling from high places. Second, it wasn't operationally robust. Our prototype would easily be stymied by uneven surfaces or small objects left on the floor. Third, Scamp wasn't economical. The components we'd used to accelerate prototyping were too expensive for the final version. Fourth, we hadn't worried about complying with consumer safety standards. Finally, and importantly, our hand-crafted artifact was not suitable for mass production. We needed to rethink the entire design to make it compatible with modern, low-cost fabrication methods.

Accomplishing all of this would take a great deal of work—many times more work than was needed to build Scamp. Development of DustPuppy would turn out to keep five full-time engineers, a manager, an assistant, and various temporary helpers and consultants busy for nearly three years. Once again, we didn't have enough cash to foot the bill for the whole shebang. And so, again, we needed a partner.

DOI: 10.1201/9781003540489-9

PROCTER & GAMBLE

This time, we thought we could find a willing confederate. In the eight years since Bissell and GoldStar had yawned at our suggested collaboration, iRobot had built a reputation in the industry and developed relationships with a few big companies. So, when Colin began phoning for dates, he attracted the interest of two potential partners: SC Johnson (our benefactor from the late Clean project) and Procter & Gamble (the multibillion-dollar consumer goods company). Both offered attractive benefits.

Each company exercised diligence—demanding that we show them what we had and detail what we planned to do. This we accomplished through meetings, documents, and slides. But Procter & Gamble would add one more hurdle. It turned out that an internal group at P&G had already started developing a floor-cleaning robot. The company might proceed on their solitary path, or they might engage iRobot to do the work. Which way to go?

Balancing competing interests, P&G management decided that choosing the optimal path required a decisive contest: single combat. Two robots, girded for action, would meet on the field of battle (a suburban Cincinnati living-room floor). There they would fight to the death and to the victor would go the spoils.

Well, somewhat less dramatically, the robots wouldn't actually meet. Rather, a representative from P&G's internal technical group and a representative from ours (me) would separately demonstrate our robots. To eliminate the possibility that one group might copy the ideas of the other, the technical people would not be allowed to see each other's robot. Only the non-technical executives would observe both performances.

On the appointed day we arrived at a pleasantly modern home near Cincinnati. Logistically, the arrangement was that the Court of Honor—Colin and Winston along with executives from P&G—would gather upstairs. Then one technical representative of each company would present their robot while the other holed up in the basement. I sallied first in our serial joust. Like every demo, this one began with me quickly surveying the room for anything that might trouble the robot. Finding none, I placed DustPuppy on the floor, switched the power on, and pressed the start button.

DustPuppy delivered a typical performance—driving about, turning away from obstacles, and cleaning bits of dirt from the floor. No surprises.

Satisfied, I headed down the stairs to the basement where my counterpart from P&G was waiting. I was disappointed that, in contrast to the usual comradery within the small robotics community, as we passed my rival maintained an expressionless, straight-ahead stare. Minutes later, I heard the faint bumps and clicks produced by his robot though could discern but little. As with my run, his seemed to have proceeded glitch-free.

But there was one glitch that day—it lay in the plan for strictly separating technical and non-technical people. Oddly, P&G considered neither Colin (who had built many robots) nor Winston (who understood physics at a PhD level) to be "technical." Following the two demos, our executive duo described to me what they'd seen. Earlier we'd heard a rumor about the nature of P&G's robot, but I was still amazed when Colin and Winston gave their account.

As mentioned previously, my AI Lab colleague Anita Flynn and I wrote a book (*Mobile Robots*) in which we instructed readers on how they could design and build their own robot. Included in our work were plans for an example robot, later turned into a kit, and marketed as Rug Warrior (named after my Talent Show robot). Astonishingly, the robot my P&G counterpart demonstrated at our almost-cordial death match was a Rug Warrior. The sole physical modifications the P&G group made were to add decorative ears and an attachment that dragged around a cleaning cloth.

Perhaps my counterpart's steely, determined focus carried the day for P&G's internal group as we did not win the competition. Our loss came despite having designed both of the robots in the contest. P&G decided to tap their internal group to develop their robot floor cleaner. But the company took steps to insulate their brand. They formed a spinoff called "Convenience Companions." With the motto "Dedicated to bringing robotics home," Convenience Companions imagined robots that would perform not just floor cleaning but other duties as well. These might include home surveillance, entertainment, and aromatherapy. Unfortunately, Convenience Companions followed the paved-for-heavy-traffic highway to robot oblivion. No products emerged from the startup's hopeful beginnings.[1]

SCJ REDUX

While courting P&G, we continued to engage with SC Johnson. SCJ required no similar formal competition, but they did insist on seeing how our robot would perform in an actual home. Thus, we journeyed to Chicago to demonstrate a DustPuppy prototype in the home of an SCJ executive.

The dwelling was equipped with an attractive hardwood floor. But unknown to the owner (or to us) in a spot near a reclining chair one of the nails securing a floor plank had worked its way upward. The nail's metal shaft was left protruding about a quarter inch above the surface. When DustPuppy rolled over it during the demo, the nail snared the robot, halting it in mid-clean. This required that I free the robot so it could continue its work, but that event was the demo's only flaw.

Ultimately, the nail incident was forgiven, and SCJ agreed to supply much of the initial cash and marketing intelligence needed to develop DustPuppy. We were thrilled to be funded but it came with a catch. SC Johnson, like P&G, was in the *consumables* business.[2] Consumables was an attractive business model because, once you've convinced customers that your product is good, they keep paying you for it again and again. It seemed natural to SCJ that the robot should fit into this, their standard business practice.

Thus did those paying the bills decree, "DustPuppy shall have a consumable." That is, some part of DustPuppy must be used up during operation so customers will keep paying SCJ to use the robot. This didn't match our parsimony-focused ethos. For a robot to sweep or vacuum a floor neither liquids nor sprays nor powders nor pads are required. All the robot needed to do was move dirt from the floor into a holding chamber. When full, the customer would empty the chamber.

As our team saw it, insisting on a consumable for DustPuppy made as much sense as applying autocorrect to Jabberwocky. Nonetheless, we were tasked with devising a consumable for DustPuppy. A cardboard box analog of the disposable dust collection bag found in most manual vacuum cleaners provided the quid pro quo required to start DustPuppy development.[3]

NOTES

1 They did, however, patent some aspects of their concept: US6459955B1.

2 A consumables business is one whose products are designed to be consumed. Such businesses manufacture items like food, office supplies, cleaning supplies, and so on.

3 A standard vacuum needs a bag—the bag acts as a filter that prevents dirt collected from the floor from flowing into the vacuum's impeller and motor. By contrast, the brushes of a carpet sweeper force dirt directly into the dust collection chamber. A disposable dust box serves no essential purpose.

CHAPTER 10

Trilobite

A N OCEAN AWAY FROM Somerville, Massachusetts yet another company had heard the sirens sing of robot vacuum cleaners.

Per Ljunggren was skeptical. The director of the New Products division at Electrolux (an appliance manufacturing giant based in Stockholm, Sweden) had just tapped Per, a veteran project leader, to spearhead the revolution. His new mission: Develop the world's first robot vacuum cleaner for consumers. The benefits of success were clear to Per, but so were the obstacles in his way. Per just couldn't convince himself that *any* path led to a viable product—the cost of the technology was sure to make the retail price prohibitive. But his boss brooked no demurral. "If anyone can do it, it's you!" Per's director reassured him.

Thus did a heavyweight contender for the robot vacuum championship step into the same ring as featherweight DustPuppy. In one corner we find shrewd corporate strategy, copious resources, and a world-renowned brand. In the other, technical inspiration, parsimony, and complete freedom to innovate. Touch bumpers and may the best bot win.

SAGA

Despite Per's hesitation, the director of New Products had a compelling reason to believe this was a risk worth taking for Electrolux. Why? Because Electrolux knew vacuums. Founded in 1919 by vacuum cleaner salesman Axel Wenner-Gren, Electrolux's first vacuum incorporated an innovative, easy-to-pull runner design that swept the market and quickly established Electrolux as a global player. In addition to their very successful line of vacuum cleaners (every fifth vacuum sold in the world was an Electrolux[1]), the company had branched out, in a big way, into other appliances. Tens of millions of their products were purchased each year giving the company an annual income of around about $10 billion.

Hans Werthén, the storied Electrolux CEO, had greatly expanded Electrolux's fortunes during his recent tenure. He'd bought up many smaller, competing companies, turning Electrolux into a powerhouse with a broad technology portfolio. Starting in the 1970s, Electrolux pioneered the use of industrial robots for the manufacture of their products.[2] Doubling down on automation, the company also became a sales agent for Unimation. Unimation, the first company to market industrial robots, was founded by trailblazing robotics entrepreneur Joseph Engelberger.

In 1983 Westinghouse acquired Unimation. The move left visionary Engelberger chafing at his new boss's glacial pace. So, he left to start a second pioneering robotics firm, Transitions Research Corporation (TRC). His new idea was to build *service robots*.[3] Early on, in a mirror of iRobot's dialing for dollars, Engelberger contacted Electrolux to inquire whether there were possibilities for cooperation.[4] The two companies considered several potential product developments. These included a robotic lawnmower, a small robotic vacuum, and a robotic autoscrubber (like iRobot's Clean machine) for commercial buildings.

Ultimately, TRC and Electrolux settled on two ideas that they would pursue vigorously. One was a robotic vacuum cleaner to be used primarily by housekeeping staff to clean hotel rooms, the second was a robotic autoscrubber for schools, hospitals, and other large buildings. To facilitate progress, Electrolux sent two engineers and $10 million to TRC headquarters in Danbury, Connecticut.[5]

The vacuum cleaner was called Star.[6] Star was similar to the Sanyo robot described earlier in that the TRC machine was also powered via an electrical cord plugged into a wall socket. But Star covered the floor in a series of star-shaped patterns as opposed to Sanyo's more prosaic rectangular motif. The patent for TRC's machine includes a figure that shows how the robot was intended to move—in the absence of dead reckoning errors. But in practice, Star would have encountered the same problems of positional uncertainty and cord tangling as Sanyo. Star never became a product.

For their industrial autoscrubber, TRC decided to base their robot on a manual autoscrubber made by the Kent Corporation, a US manufacturer of commercial cleaning equipment that had been acquired by Electrolux in 1969. The developers would keep the basic structure of the machine, including its manual controls (so a worker could operate it manually if needed) but they would add sensors and a microprocessor to enable robotic control. As with the square-shaped Zoom Broom from the AI Olympics, this choice meant that no work would be needed to achieve good cleaning but solving the robotic problems would become more challenging.

This led to uncomfortable product performance constraints. To make navigation (one of those robotic problems) tractable, developers stipulated that the robot would operate autonomously only in long corridors. Any physical geometry more complex than a simple rectangle would confuse the robot. In the era before SLAM solved the indoor robot localization problem, this restriction made the robotic autoscrubber technically and economically viable. But only just.

After many years of development, first at TRC and then at Kent headquarters in Elkhart, Indiana, the new robot, designated RoboKent, was offered for sale. The product survived but did not thrive. Because of its added robotic elements and limited sales volume, RoboKent was more expensive than manual autoscrubbers with the same capacity. Another drawback was the need to train operators to use the unfamiliar equipment. Such training was a constant challenge for commercial cleaning operations where worker turnover was very high.

ROAD TO DAMASCUS

Electrolux's experiences with Star and RoboKent tweaked Per's unease. The meager return from the company's earlier robotic cleaning efforts was not commensurate with the large investment they'd made. Was it truly possible to do any better, he wondered? Regardless, Per had his orders. So, with reluctance, he accepted the mission his director had given him and took up the mantle as project leader for the new home-cleaning robot.

What was the proper way to build a robot vacuum for consumers? In a corporate move that those of us working for minimally funded startups could only view with rueful envy, Per decided it was best to explore multiple solutions simultaneously. He recognized navigation as a central challenge, so he solicited proposals from research departments at several different university and consulting organizations. Two engaged and were given funds to develop an approach; a third solution was attempted by an internal group at Electrolux. Approximately a year and a half later all three parties demonstrated working robots.

One group showed off an impressively sophisticated system that reliably kept track of the robot's location. Centrally mounted on the robot was an ultrasonic emitter. In the same room as the robot were several beacons. Each beacon contained an ultrasonic detector and a radio transmitter. When a beacon detected the ping from the robot it responded by immediately transmitting its unique ID number over its radio channel. When the robot then received a radio transmission from a beacon, the robot saved the ID and checked its timer. Neglecting the speed of light,[7] the current value of the timer represented the time taken by the sonar ping to travel from the robot to the beacon.[8] The robot could then multiply this number by the speed of sound to compute the distance between itself and each beacon. If there were at least three beacons in each room, the robot could use their relative distances to compute its location within the room.[9]

Watching this robot work was to behold a thing of beauty. Because it always knew exactly where it was, the robot could navigate flawlessly. The robot could cover the entire room, missing nothing. The second group delivered a system that, like DustPuppy, relied on random bounce. This robot never knew where it was but nevertheless also succeeded in covering the room based on the mathematics of chance. The third group proposed a hybrid approach that combined some knowledge of position with random bounce.

The last method didn't seem to add anything new technologically. So Per scrapped that approach but he arranged to have the other two robots tested extensively. It turned out that in a realistically cluttered room, the difference between precise navigation and random bounce didn't amount to much.

That was a good thing because, in Per's estimation, adding the cost of the precise navigation beacons would push the price of the robot over the border into nobody-will-buy-it land. Later he would decide to ditch the beacons but add the sonar transducer to the random-bounce robot. That would enable the robot to track very short distances, ensuring that it would lightly kiss walls and other obstacles rather than slam into them. The sensitive sonar could even detect and avoid thin power cords dangling from tabletops![10]

The robot had proven itself. It covered the room, it would accommodate a cleaning mechanism, and, Per computed, it ought to be sellable at a nonprohibitive price (at least in Europe). Like Saul traveling to Damascus, Per saw the light—his skepticism vanished in the manner of dirt from beneath a robot vacuum. He truly believed that Electrolux could be the first to market a viable robot vacuum cleaner for consumers. The newly minted messiah of clean now needed to assemble his disciples.

FROM TRIBULATION TO TRIUMPH

The New Products group was set up to foster internal entrepreneurship—a difficult proposition for any mature company. New Products would provide seed funding and, at least initially, freedom from the many strictures that accompanied typical, incremental products. Ultimately however, the sheltered neophyte would come to market only if it could gain admission into one of Electrolux's existing product lines.

Per was granted the liberty to pick whomever he though best suited to the development task—including engineers not employed by Electrolux. Per opened his extensive contact list of creative, highly capable engineers and started dialing. Soon, a hand-picked core team of seven engineers, about half from outside Electrolux, set to work. In short order they had invented and began patenting[11] key methods the new robot would need. The technical side of the development proceeded smoothly and within 18 months the team had a version of the robot that was able to navigate and clean.

The first working prototype "looked like a giant cookie tin" according to one engineer. Those of us in the business understand that this is an entirely appropriate look for an engineering prototype—functionality first, style later. But it would create problems for Per.

The appropriate internal group to bring Per's robot to market was the Floor-care department. It was their favor that he had to curry. But he expected that would be easily accomplished at Floor-care's international sales conference held annually in Stockholm.

At the critical conclave Per presented his masterpiece-in-progress. From a technologist's perspective, the robot's tremendous potential was self-evident. No doubt, Per felt very much the same as Jack Shimek and I had when, with great anticipation, we demonstrated RoboBroom to our colleagues and Denning Mobile Robotics' CEO.

The outcome for Per was only moderately less disastrous.

The Floor-care department was horrified. "No one will buy this!" they declared. The robot seemed crude; it looked nothing like the refined products the division normally marketed. To protect the company from an embarrassing flop, Floor-care vociferously declined to adopt, leaving Per's robot an orphan.

Fortunately, unlike the catastrophe that befell Jack Shimek and me at Denning Mobile Robotics, Per wasn't fired. But Floor-care's rejection halted his project and scattered his team to the wind. A true believer does not accept defeat (I can attest). But temporarily retreating and regrouping seemed prudent. Maybe Per had asked too much of the imaginations of his compatriots. When Per looked at the robot, he saw the beautiful swan it would become, but Floor-care saw only the ugly duckling it was now.

For Floor-care to fall in love they needed to see the robot's inner beauty fully expressed on its exterior. Per's robot needed a makeover. So, the first priority was to gather more funds. This required considerable time and certain machinations but Per eventually secured the cash he needed to finance another run at the prize.

Next, he had to find a makeover artist. The industrial design (ID) department at Electrolux was available for this purpose. But Per saw them as too cozy with Floor-care. If the ID folks were involved, they would insist on applying their standard sensibilities and heuristics to the design. But for this product, aesthetics and functionality could not be neatly separated into silos. Unless the industrial designer internalized the intimate relationship between form and function, the product might not work properly.

Per played his ace in the hole. His wife Inese was an accomplished artist and industrial designer, and she set to work assisting the team providing her services for free. Approaching the difficult design challenge, Inese reimagined the ill-favored "cookie tin" as the sleek, modern reincarnation of a trilobite, a creature that inhabited ocean bottoms during the Cambrian period. The result of her efforts was nothing less than stunning. It won awards.[12]

The robot, now matured into a ravishing swan, assumed the name of its muse, Trilobite.

Per had ten copies of this beautiful version of the robot hand built. These he loaned out to key constituents to gain practical experience and to build support. Per planned to propose the now exquisite Trilobite again at the next Floor-care sales conference. But fate intervened.

A film production crew from the BBC was visiting Sweden. They had arranged with ten local companies to record segments showcasing each firm's advanced technology. The stories would air on *Tomorrow's World*, the BBC's forward-looking documentary series. But one of the featured participants unexpectedly dropped out. Seeking a last-minute substitute, a BBC representative called Electrolux and asked, "Do you have anything interesting we can film?"

The request found its way to Per, with whom it sparked a dilemma. During the several years the project had been underway, Per and others had made heroic efforts to ensure that no hint of Electrolux's robot vacuum project leaked out. They had succeeded. Was it now time to abandon secrecy and trumpet the news to the world? Might that advantage potential competitors? As a result of his collision with Floor-care, Per had reached a conclusion. Given how difficult it was to convince his own colleagues at Electrolux that a robot vacuum cleaner was a good idea, it would likely be even tougher to convince the rest of the world. Educating the market would take time but no robot could succeed without it.

The BBC's offer seemed like a golden (and free!) opportunity to begin that process. As soon as Per showed the BBC crew just how interesting Electrolux's "interesting thing" was, they decided to reset. Trilobite merited something of much greater impact than a filmed-on-location, one-of-ten technology story. Production would be moved to the BBC's London studios where they could give Trilobite the VIP (VIR?) treatment.

And so it was that on May 10, 1996, on an episode of *Tomorrow's World*,[13,14] Electrolux broke the news. Presenter Philippa Forrester, dressed mostly in a towel (for reasons known only to the BBC producers), demonstrated the amazing abilities of the elegant new Electrolux Trilobite robot vacuum.

For its TV debut, Trilobite was simply gorgeous. Its visage may have been inspired by an aquatic creature, extinct for a quarter billion years, but the robot's appearance was in every way sleek and contemporary. Round, standing a little over five inches tall, possessed of a dark green, mirror-like finish, Trilobite glided smartly about the floor. It delicately touched then turned away from a full glass of wine inadvisedly perched on the carpet, it scooped up debris, it maneuvered in near silence—a robotic tour de force!

Left unsaid on the broadcast was what Trilobite might cost or when it might be available.

***

In 1996, three years before conceiving DustPuppy, we at iRobot did not catch the *Tomorrow's World* broadcast. We did, however, encounter several news stories reporting on Electrolux's amazing robot. As a company of robot enthusiasts, we found this most exciting. And, despite all the earlier false starts, we dared to hope that a corporation of Electrolux's stature and expertise might be poised to splinter the logjam, finally making good on the promise of robot vacuums. Our whole industry, we thought, would benefit from such a success.

But nearly four years later, as the DustPuppy project started up, we'd heard no further updates from Electrolux. As a result, I was persuaded that Trilobite had copied the same forlorn formula implemented by all previous would-be robot revolutionaries. First excited hype, then silence, and next a slow, wistful fade into oblivion. But this time my conclusion was hasty; the surprising truth would soon visit terror upon the DustPuppy team.

NOTES

1 Electrolux press release, 1998: https://www3.electrolux.se/corporate/pressnotes/42c2_4b2.htm.
2 Electrolux Annual Report 1975, p. 6.
3 Rather than assemble things, service robots provide a service like cleaning floors or moving items in a warehouse.
4 A. Bancroft, "The First Commercial Floor Care Company that Ventured into the Production of Robots," *Conference on Intelligent Robotics in Field, Factory, Service and Space*, Volume 2, (CIRFFSS 1994), https://archive.org/details/nasa_techdoc_19950005096/page/n229/mode/2up.
5 B. Ward, The Coming Tsunami: Service Robots, https://cybercleansystems.com/TheComingTsunamServiceRobotsv2.pdf.
6 TRC's robot vacuum patent, US4962453A, was granted in 1989.
7 It makes sense for the robot to ignore the speed of light in this calculation because light travels about one million times faster than sound. Taking light speed into account, accuracy would improve by only 0.0001 percent.
8 Of much greater consequence than the light travel time are delays due to the processing electronics in the beacon and robot. Ideally, such delays are constant and can be removed from the distance computation.
9 A patent was granted for this system: EP0753160B1.
10 We worried that DustPuppy might cause trouble in such a situation. If it encountered say, a cord supplying power to a table lamp, the robot might run down the cord and pull the lamp off the table. But adding sensors to eliminate this rare hazard would increase DustPuppy's price. So we decided to make the user responsible for repositioning dangling cords.
11 European patent: EP0841868B1.
12 Inese Ljunggren's design (US Patent D375592) contributed to the robot being selected as the "the most exciting design product of 1997" in the book *100 Designs, 100 Years: Innovative Designs of the 20th Century* (1999), ISBN 978-2880464424. The robot was also exhibited in the National Museum of Fine Art in Stockholm.
13 https://genome.ch.bbc.co.uk/schedules/service_bbc_one_london/1996-05-10.
14 Video: https://www.youtube.com/watch?v=E0hkGUCmxy0.

CHAPTER 11

Base Camp

W HEN SCJ PULLED THE plug on Clean, the project entered a winding-down phase. And while most team members moved to new projects immediately, Winston and a couple of others—one being Phil Mass—stayed on for a time. Phil and a colleague continued to develop some of the navigation work we'd started, and they undertook an orderly mothballing of the project.

This put Winston and Phil in close collaboration, enabling Phil to learn early details about the new DustPuppy project Winston would be managing. Phil was a long and strong proponent of the minimalist, bell and whistle-free DustPuppy ethos. He'd been planning to apply those principles to a project that used a large swarm of robots to accomplish tasks, but Winston convinced Phil to join him on DustPuppy. As long as the new project worked to avoid the pitfalls of Clean, Phil was in.

I was excited to welcome Phil to the team when he joined in March 2000.

GALE WARNING

As we assembled our team and ramped up the DustPuppy project, we were excited, eager, and fully confident of success. For over ten years I had been learning and gaining experience, working toward just this sort of opportunity. My colleagues were talented, hard-working, and conscientious. What could stop us now? The question was not rhetorical, as formidable forces were arrayed against us.

First, history was in robust opposition. As we've seen, regardless of how much our robot floor-cleaning predecessors had invested, they all ended up with a return of zero. What sort of responsible company management would commit millions of dollars to an enterprise that, given the historical 100 percent failure rate, had a negligible chance of success? Fortunately, at that time no head-examining resources were available on staff at iRobot.

Operating in the real world our robot would constantly meet surprises. That's because, unlike Handey, it would never know its exact situation. To compute its next move, DustPuppy would rely on the limited information about its environment provided by the sensors we could afford to include. Every region, home, and room present unique challenges, yet we wouldn't be able to anticipate the specifics of each trial DustPuppy would face. Regardless, we had to manage them. The ace up our sleeve, behavior-based programming, wouldn't magically eliminate surprise; it just provided a powerful tool to help us deal with novelty. However, owning a hammer doesn't guarantee the ability to construct a house. Would we truly be able to create code that handled all situations? Programming our robot would pose one of our greatest risks.

The technical challenge of designing robot hardware was the peril that worried us least. But realistically, it wasn't going to be easy either. Matching the fictional robot in Heinlein's book *The Door into Summer*, our aim was to sell a robot cleaner at a manual-vacuum price. That put a lot of standard technical solutions—popular in academic papers on robotics—out of reach for us because they cost too much. Could we invent alternatives with more comfortable prices? I was certain we could, but I'd been diagnosed with chronic optimism.

We teammates believed we had an exquisite command of our domain because we all had extensive experience cleaning our own floors. But we were also technologists who loved robots—in contrast to our typical customer. Would our intuition guide us reliably or lead us astray?

Had we truly put a stake through the heart of the Clean project? Might the fiends of acrimony and disunity scrabble out of their grave to torment DustPuppy? None of us foresaw that happening, but then we were surprised by Clean's descent into dysfunction.

The bottom of the corporate ocean is choked with projects sunk by company politics. We believed Winston would act as our sturdy bulwark against interference from the outside. But there were no sure things.

Competition was a "known unknown." A serious competitor could emerge, but that sector of our danger chart was initially labeled TBD. We most feared deep-pocketed rivals. Developing DustPuppy would be close to the limit of what iRobot could pull off. A serious challenge from a major player could easily have exceeded our ability to respond. We comforted ourselves with the knowledge that all the folks currently developing robot floor cleaners[1] were doing it wrong (as we saw it). But that could have changed at any moment.

Finally, there could have been problems that we hadn't identified as problems, the proverbial "unknown unknowns." We could identify some of the punishers lining both sides of the gauntlet before us. But there could have been others we didn't know, invisible adversaries that could have struck and sunk us without a trace.

Despite our sanguine confidence, DustPuppy was a big risk.

OFFICE

During DustPuppy development iRobot's offices occupied an unlikely venue. Technology companies typically situate themselves within office parks, cheek by jowl with compatriots of similar persuasions. Not iRobot. We were the proud, if lonely, high-tech tenants of a well-worn shopping center in Somerville, Massachusetts known as the Twin City Plaza. (Cambridge being our shinier sibling city.)

Entrance to the iRobot offices required climbing the stairs to overtop the first-floor shops and walking past our neighbor, MINE (Massage Institute of New England). Our headquarters consisted of a labyrinth of offices, corridors, and workspaces, including a machine shop, a freight elevator, and a giant room we called the "high bay." The only windows in our space looked out toward the bleak, seven-story-tall cold storage warehouse that was the nearest neighboring building. But we weren't there to look out the window.

That was fortunate because the DustPuppy office had none. But above the hallway outside our door was a small skylight. This sometimes allowed in a ray of sunlight in to cheer our 20-foot square workspace with cubicles around the perimeter. The initial inhabitants were Paul, Phil, and me. Winston maintained his marginally more prestigious office in another part of the building. At the center of our room was an elegant, modern office table. When we placed it there our plan was to hold team meetings around the table. This pristine surface was rapidly covered with a large sheet of brown paper so as to protect it from the robots, robot subassemblies, and various other components that took up permanent residence— the demands of hardware were more important than the convenience of meetings in our group.

Over time, as the need arose, we built and dismantled various architectural features in our space. A robot "pen" formed from two-by-six-inch lumber occupied the space just outside one cubicle. During Winston's visits to the team he seemed irresistibly drawn to the challenge of walking along the narrow tops of these boards. Inevitably he would take a step or two, flail wildly, and then step onto the floor of the pen to regain his balance. Despite near daily practice he never got any better! Happily, Winston never failed at a much more important balancing act—keeping the Dust Puppy team focused and unharmed by the political caprice of two separate companies: iRobot and SCJ.

We had our space but our team remained incomplete. We needed champions to design the production versions of the electrical and mechanical systems DustPuppy must have to become a real product.

CHRIS CASEY

The TRS-80 was an early personal computer. Low in cost, modest in features, and sometimes derisively dubbed the "Trash-80," this machine served as the introduction to computing for many future engineers. Chris Casey first encountered one at a summer camp in Oregon when he was 14. Never having contemplated computers before, Chris was instantly fascinated. With a computer as gateway, he spent the summer exploring an unexpected and wondrous new world.

On his return home Chris initiated a lobbying campaign. Though his parents were of humble means, Chris labored to persuade them that the entirety of his future success hinged on having his own TRS-80. On Christmas day the new computer arrived, and Chris departed. Holed up in his room over the next few months he apprenticed himself to the machine. First, Chris mastered the computer language Basic. Then, diving deeper, he became proficient in assembly language, the arcane instructions that underpinned computer languages and granted full power over the machine.

Software was fun but to Chris's mind incomplete. The programs he wrote treated the computer as a black box. But what internal magic transformed keypresses into glowing pixels on the screen? Maybe he could figure that out if he built his own computer.

Honing the persuasive skills that had earned him his TRS-80, Chris contacted manufacturers of electronic components. He requested free samples of items that might be useful in building a computer. Many companies obliged. But Chris's procurement turned out to be premature. As a high school student he lacked a few bits (ahem) of crucial knowledge needed to convert his ideas into working electronics. But his passion was confirmed, and electrical engineering became his destiny.

Chris followed that track through college at MIT, graduating as a bona fide electrical engineer. Then Chris discovered one more passion: gambling. By virtue of a friend's bachelor party at a casino, Chris encountered blackjack. Many gamble for the thrill of casting their fate to the wind, winning or losing on the turn of a card, but not Chris. The element of blackjack that most attracted him was statistics. The outcome of any single game of chance is unknowable. But the aggregate outcome of many games was an absolute certainty as the laws of probability were obeyed without exception. Understanding and exploiting those laws gave Chris an edge. Declining to play except when the odds were in his favor meant that the risks Chris took ultimately always paid off.

Risk brought Chris to iRobot. Before children, mortgages, and responsibilities mandated prudent rather than exciting choices, Chris decided to take a risk on a company that had unicorn potential. In his mind, the outcome of that single throw of the dice would depend on luck alone, as it would for any reckless gambler. In reality, Chris's considerable talents would serve as his edge, tilting the odds in favor of both him and his new company.

Chris began his employment at iRobot two years before DustPuppy's initiation. During the biennium he lent his expertise to several robot projects. But the alliances were fraught.

It happened semi-regularly. At weekly meetings of iRobot's project managers, manager A would say: "My project needs another electrical engineer."

Manager B: "Chris Casey is available."
Manager A: "Right. But I really need another EE—who can I have?"

Chris had developed a reputation for being difficult. When managers chose their teams for a game of project development, Chris tended to be picked last. Winston, attending those manager meetings, had absorbed the conventional wisdom. So, he felt considerable unease when, near the start of DustPuppy, Colin told Winston that Chris Casey would be our electrical engineer. To judge whether or how hard to push back, Winston consulted our human truth sayer, Phil. Phil said, "Chris will be fine."

Phil had intuited that Chris's "difficult" wasn't the same sort of "difficult" that some members of the Clean team had practiced. Chris's focus was always on the project and the execution of *all* the steps needed to complete it, and he would not allow anyone to put off or avoid addressing hard problems. Chris could poke holes in any rosy projection of how things were going to go, and he seemed to delight in identifying the many ways a project or design might fail—whether he was invited to weigh in or not.

Deference to hierarchy is standard protocol in the corporate world and in typical product development. But when breaking totally new ground, the boss possesses no monopoly on guiding insights. In that case a willingness to challenging orthodoxy, pet ideas, and the team's own Kool-Aid is essential to success. This was where Chris' relentless focus would prove decisive.

DIMENSIONS

How big should DustPuppy be?

Ideally, we needed a robot that channeled Doctor Who's Tardis[2]—small on the outside but much larger within. DustPuppy required a diminutive external stature so that it could crawl beneath couches and beds and kitchen toe kicks[3] to gather the dirt that accumulated there. Also, to facilitate maneuvering betwixt the legs of stools, dining room chairs, and such, the robot should be slim. But, at the same time, our machine ought to be of ample size so as to accommodate an effective cleaning mechanism, long-lasting battery, generous dust receptacle, and all the other hardware that covering a large living room would demand.[4]

Without access to the Doctor's dimensionally transcendental technology, we were forced to seek a compromise between large and small. But choosing the robot's dimensions carried substantial consequences. Published standards might have decided the matter for us—sadly, there were none. Manufacturers were free to build furniture with as much or little clearance underneath as they chose. And kitchen designers could specify kicks fit to any hypothetical toe. But we did have access to one set of useful statistics.

Our partner, SC Johnson, had collected vast amounts of data about their customers' homes and possessions, including things like the range of under-furniture heights and toe-kick sizes. They shared their knowledge with Winston—and he agonized over it. As he made a theoretical robot taller and wider he found fewer and fewer places it could visit. SCJ's data showed that, depending on the starting point, adding a quarter inch of height, say, might eliminate a million potential customers. Clearly, a good fit of our product to the marketplace demanded the stubbiest of robots.

On the other hand, the shorter the robot the more extreme the migraines our mechanical engineer would experience trying to squeeze all the essential components into the available volume. At some non-obvious minimum such efforts would shift from heroic to hopeless.

Winston found his attempts to compute the optimal compromise excruciating. He well understood the unhappy market implications of building a too-tall robot but the engineering realities that would render a too-short robot unworkable escaped him. He needed additional expertise to balance marketing needs against achievable technical specifications. Happily, just such an ability turned out to be another of Phil Mass's numerous talents. As we'll soon see, Winston and Phil spent many hours contemplating this and numerous other issues. They knew the height challenge would be tough.

To achieve a robot outline that was small, but not too small, the team needed a mechanical engineer both fearless and stoic enough to push the envelope separating merely punishing from outright impossible.

ELIOT MACK

Eliot Mack was born in Nebraska under the wide-open skies of America's heartland. As he grew, he discovered within the pages of *Popular Mechanics* and other heralds of the future fascinating, compelling stories of flying cars, thinking machines, and other high-tech wonders. Dismayingly, none of these exciting stories were set in his hometown of North Platte. That ignited in Eliot a growing desire to go where the future was being forged.

In early preparation Eliot took up computers, programming a series of Radio Shack, Texas Instruments, and Apple computers. But his enchantment with technology extended to machines of all sorts, especially cars. Much of his childhood was spent constructing one project or another. Disconcertingly, disaster seemed to stalk his efforts.

Highlights (or lowlights) included accidentally catching his prized 1972 VW Beetle on fire while, ironically, welding on a piece of safety equipment. In an earlier episode, he built a gravity-powered go-kart. Unfortunately, Eliot installed the axle threaded for the right side on the left side. Part way down the hill, the left front wheel abruptly departed. This distressed younger brother Adam whom Eliot had recruited as test pilot. (Decades later, Adam continued to cast doubt on Eliot's mechanical designs, especially ones that involved wheels.)

In high school Eliot first encountered *audio animatronics*[5] in the form of a talking buffalo head. The bovine simulant was mounted to the wall of the Fort Cody Trading Post store where Eliot worked during the summers. Eliot's job was to initiate a performance every half an hour, which enabled Cody to serenade customers. (Yes, the buffalo head had a name.) Standing at his post beneath the talking buffalo head Eliot glimpsed an unsettling image of his possible future.

The bodiless creature proved to be a potent motivator. When he arrived at the University of Michigan, Eliot was haunted by a specter—unless he excelled at his studies (thus attaining North Platte escape velocity), a lifetime in service to the talking buffalo would be his destiny.

Eliot did well in college.

He went on to graduate school at MIT and from there to a job at Product Genesis, a nearby product design company. Eliot spent an intense few months completing his assignment to design an electronic toy—a virtual fishing rod. The product involved an electric motor, electronic controls, and an injection-molded plastic case. The client demanded high functionality at the lowest possible cost. This job, lasting just over a year, was pivotal. During this time Eliot learned the critical product engineering and injection-molding design skills that would later shape Roomba.

Next stop was the newly established Cambridge office of Walt Disney Imagineering. The exclusive focus of Eliot's group was designing a 13-foot tall, 11,000-pound walking, emoting, robotic triceratops.[6] Disney executives planned to one day release the dinosaur into one of their theme parks. (What could go wrong?)

The dinosaur project was high-profile, highly funded, and highly monitored. Eliot discovered that abundant money and attention were not always beneficial to a project. Clashes developed between the artistic leadership in California and the engineers in Cambridge. The artistic folks back in the Golden State had developed processes that worked well in their familiar domain. The engineers could see that applying those methods to an autonomous robotic dinosaur was not a recipe for success. But they could not persuade the artists. After several frustrating years the project fell apart.

It was at this point that Eliot decided to follow a couple of friends who'd earlier left Disney in favor of jobs at iRobot. Eliot applied and ended up being interviewed by Winston and Phil. His experiences and qualifications made him stand out as a candidate for DustPuppy.

The trio engaged in a long conversation, most of it taking place outside as they strolled around the Twin City Plaza. During the walk, Winston and Phil warmed to the Nebraskan, concluding that Eliot's mix of enthusiasm and technical skill would indeed be a great fit. In talking up the project they showed Eliot the Scamp prototype. Regarding it nonchalantly, he thought to himself, "It's just three motors; how hard could that be?" (He'd have ample cause later to reassess that notion.)

Later, while settling in, Eliot pondered further whether the DustPuppy concept was viable, and whether the engineers on his team were equal to the challenge. Phil's quiet competence was apparent to him, and his initial take on Chris Casey was that any electrical engineer that ornery was likely to generate circuit boards that worked.

NOTES

1 Besides Electrolux, we knew of several contemporaneous projects. They included: Kärcher, a German conglomerate; Dyson, the British vacuuming powerhouse; Friendly Robotics, an Israeli company; and Floorbotics, an Australian startup.

2 Doctor Who, of the planet Gallifrey, zips about the universe in a spaceship/time machine called the Tardis. Notoriously, the Doctor's conveyance is much bigger on the inside than on the outside.

3 A toe kick is the recessed area typically found under kitchen and other cabinets. The toe kick accommodates your toes allowing you to comfortably stand closer to the counter.

4 Patents reveal that our predecessors had chosen mostly tall machines. That simplified their mechanical engineering challenge but ceded the area under beds and toe kicks to the dust bunnies.

5 Unlike this author, many consider such constructs to be robots.

6 The Disney dinosaur is described here: https://www.discovermagazine.com/technology/a-giant-among-robots.

CHAPTER 12

Children of Necessity

W ITH THE DUSTPUPPY TECHNICAL team complete, we accelerated to full development speed. Implementing our robot at an attractive price required that we devise several enabling inventions. Most were formulated during our first hectic year. (Subsequent years became even more hectic.)

LIVINGROOM LILLIPUTIAN

Imagine suddenly shrinking to a height of three and half inches. And then having to clean the carpet.

As mentioned, Winston had learned from SCJ's consumer data that our robot needed to be runt-of-the-litter short if it was to attend to all the areas we wanted it to clean. After much consideration and several design forays, Eliot and Phil concluded that about three and half inches was the minimum height we could make DustPuppy while maintaining the ability to pack in all the needed components.[1] And, the robot would need to be about 14 inches in diameter.

With the dimensions settled we were left with the equally daunting challenge of maneuvering our petite machine such that it would never get stuck. Why is that so hard? Try looking at it from robot's point of view. Literally. Get down on the floor and, as I did, position your eye a few inches above the surface. I found that from this new perspective, familiar scenes took on alarming aspects. Power cords loomed, threatening to ensnare and entangle, the odd sock presented a pernicious trap, and climbing from a hardwood floor onto a braided rug looked decidedly chancy.

We referred to DustPuppy's ability to cope with every such challenge as *autonomy*. Maximizing that quantity in every domestic environment was

 DOI: 10.1201/9781003540489-12

both critical and complicated. One facet of autonomy that generated much argument was *clearance*—the distance between the bottom of DustPuppy's bumper and the ground.

Clearance sorted the world of objects into one of two categories, those tall enough to touch the bumper and those short enough to slide under it. Objects in each class affected the robot differently. DustPuppy turned away from taller items that touched the bumper. All shorter items ducked under the bumper and encountered DustPuppy's cleaning mechanism. It would try to pick them up.

How high or low should we set the clearance?

Over the course of several months, as we worked on other aspects of the robot, we built prototypes with different bumper clearance heights. Observing DustPuppy in action convinced us to raise the clearance higher than felt entirely comfortable. This was mostly because of braided rugs; even if you started DustPuppy atop a rug, it would soon climb down never to return. We were compelled to set the clearance high enough that DustPuppy would summit all common rug thicknesses.

Solving one problem created another (a recurring theme). A high clearance meant more objects slid under the bumper. Some of them, such as socks, magazine pages, and scattered papers, could wrap around DustPuppy's brush and jam it—failing our autonomy test.

We addressed the jamming problem by adding a grille of parallel wires (we called it the *bail*) under the cleaning mechanism. The bail functioned as a sort of filter. Its wires were spaced about an inch apart—wide enough to avoid interfering with the pickup of small debris but tight enough to prevent large, mostly flat items (such as socks) from gumming up the works. Like many systems on the robot, a lot of fiddling with spacing and design was needed to make things work acceptably well.

PHASE CHANGE

Like Prometheus bringing fire to humanity, our new teammate, Eliot, brought the rarefied knowledge of injection molding to DustPuppy. His contribution transformed the project. To achieve a low retail price we needed to replicate the process Hasbro used to manufacture My Real Baby. But our familiar construction techniques weren't compatible with that method. Scamp and all our earlier prototypes had been constructed by combining a chassis made of drilled and folded sheet metal with a shell of vacuum formed plastic or cast urethane. Building such a robot was tedious; it usually required hours of work in the machine shop. Our new colleague showed us a better way.

Eliot valued the ideas embodied in our frumpish prototypes. But to realize the advantages of low-cost manufacturing, he knew that every existing component would have to be scrapped and reworked. The process consumed Eliot for the first four months of his DustPuppy tenure. He first reimagined a robot manifest in plastic rather than sheet metal and then used a computer aided design (*CAD*) program[2] to painstakingly model each part.

Once the parts were resident in the computer, we were able to take advantage of another transformative, development-accelerating technique we'd never used before, 3D printing. Now, rather than physically building every component of the robot—cutting, bending, drilling, and filing each one, one at a time—we could simply type! Write an email, attach a CAD model of the desired part, and send it off to a 3D printing firm. In a few days a perfectly formed part would show up on our doorstep. Today even humble hobbyists enjoy the advantage of desktop 3D printing, but in those days the concept was fresh and exciting. Once we banished sheet metal and urethane, DustPuppy started to feel much more real. Rather than a research project, it started to look like an actual product. But it *was* still research—we were constantly learning new things and inventing new mechanisms. Our design had to evolve rapidly to keep up; no set of printed parts was up to date for long.

When the initial main brush gearboxes were fired up for the first time, the (initially) low-precision gears howled so loudly that they could be heard outside the test room and across the office. Winston asked, in a tentative, wavering voice, "Will that get better?" We assured him it would.

A second circumstance forced us to print new versions of the robot frequently. The 3D printing process we used most often was called SLA (the contrived three-letter acronym for stereolithography). Our 3D printing vendor[3] sent us SLA parts that were perfect in every way but one—they were brittle.

The parts seemed to become even more brittle over time. When an SLA part chipped or broke, as often happened during robot testing, we would patch it back together with a combination of super glue and a glob of hot glue. (The robot obligingly helped with its own development. Whenever a piece broke off during a testing run, we would almost always find that DustPuppy had swept up the part and collected it in its dust cup.) The longer we used a robot, the more it came to resemble an amorphous blob.

Eventually, no more glue could be added; in order to continue testing, we'd have to order a new set of parts even if the design hadn't changed.

Although Eliot's Promethean gift was a critical boon, it carried a high maintenance cost. While the rest of the team argued and fretted over pennies, Eliot would be ordering—every week or two—a new build of SLA parts for $5,000 or $10,000 or sometimes $15,000. The gods demand their tribute.

SWEEPING CHANGES

Designing an effective sweeping mechanism for DustPuppy was an early order of business. We'd been appropriating and modifying the guts of Bissell carpet sweepers for long enough. Now we needed a cleaning system tailored to our robot.

Most carpet sweepers work in either direction—push the sweeper back and forth and it picks up dirt both ways. Usually, this ability implies two dust chambers, one on either side of the central brush. Debris tends to fall into one when the sweeper moves in one direction and collects in the other chamber when it moves the opposite way. Effectively, DustPuppy would move only forward. So we needed only a single dust collection chamber.

Paul and I began experimenting with a one-brush, one-collection-chamber mechanism. But we found that when our motor-powered brush encountered certain types of dirt, rather than pick it up, the brush sometimes sent the dirt shooting forward, ahead of the robot. That wasn't a good look for a cleaning robot. Also, the ejected dirt might bounce into a spot DustPuppy couldn't reach, so we sought a solution.

The simplest remedy we could think of was to install a barrier—a rigid vane hinged at the top, positioned ahead of and parallel to the brush. The bottom of the vane would touch floor but the hinge at the top let it swing out of the way of any small objects the robot ran over. The vane did indeed stop dirt from shooting out from under the robot, but it created a mobility problem. With the rigid vane in place, the robot was no longer able to reliably climb over small objects like low throw rugs that had previously given it no trouble.

So, we tried again. Maybe a flexible rubber vane would do the trick. We replaced the rigid component with a more pliant one and tried again. This too eliminated the dirt-shooting-forward problem, but it did not restore the robot's mobility.

We began to think about active solutions—this we did reluctantly because any new moving parts would add complexity to the robot.

Nevertheless, we next built what we called a flapper brush. This device was composed of six short rubber vanes protruding perpendicularly from a solid, central core. It had the same length and diameter as the cleaning brush and was positioned just ahead of it under the robot. Significantly, while the bristle brush opposed the motion of the robot, the flapper brush spun in the opposite direction and tended to advance the robot. It turned out that the net result of the two brushes improved both cleaning and the robot's mobility. That bit of complexity earned its keep.

The bristle brush copied features from typical carpet sweepers. The brush's core was composed of two stiff wires twisted around each other with thin, straight nylon bristles trapped firmly between the wires. When spun, the tips of the bristles traced out a cylinder a little under two inches in diameter; the brush was about six inches long. The springiness of the bristles was crucial, so we used the same material, diameter, and approximate length as the bristles we found in carpet sweepers.

The main brush motor powered our dual brushes. It was the biggest motor on the robot; it needed to be because it did most of the work. Spinning the brushes against a (typically) high-friction carpet took a good bit of power—although vastly less than what a vacuum would have needed. A train of several gears connected the main brush motor to the bristle and flapper brushes and made them spin in opposite directions.

Our initial tests of the cleaning mechanism were informal, involving dropping Cheerios on the ground and judging by eye how well one mechanism worked compared to a competing one. Later we did much more extensive, objective tests. We scattered a measured amount of test dirt (often sand) on the floor, let the robot pick it up, and then weighed the recovered dirt. We reached a point where one pass of our mechanism appeared to work as well as a couple of passes of a manual carpet sweeper.

A coverage test we did early on was to install our latest cleaning mechanism in the robot and let it drive around a 100 square-foot area within the DustPuppy office. Inside this square frame formed by four two-by-fours, we used masking tape to mark out a grid of squares one foot on a side. When we wanted to run a test, we would first place a small item, usually a Cheerio or tiny scrap of paper in the center of each square. We then noted how long the robot took to clear all the ersatz dirt. Due to the test area's resemblance to the *Star Trek* room of the same name (when no program was running), we called our test area the holodeck.

CHRISTENING

Although we on the technical team—unburdened by any semblance of marketing savvy—thought DustPuppy was a great robot name, others demurred. More knowledgeable authorities, Winston in particular, viewed the appellation as just the sort of name starry-eyed technophiles *would* come up with, probably inspired by a cartoon. The name was *not* so inspired though, admittedly, there was a cartoon connection.

After we'd picked the name, Paul and I learned of a character called "Dust Puppy" in the webcomic *User Friendly*. Whether that character, reputedly made of dust, lint, and quantum events, would have created trademark headaches for our DustPuppy, made of plastic and dependent upon quantum events to direct the collection of dust and lint, will remain forever untested.

Winston, assisted by Jeff Ostaszewski, our newly hired, recently graduated marketing specialist, plunged into the task of renaming DustPuppy. To help in this critical effort, they engaged Catapult Thinking, a professional branding firm. To seed the search Winston and Jeff offered some ideas. The new name shouldn't engender any sort of gee-whiz, high-tech, futuristic vibe. Rather it should evoke life, movement, and bounce.

After many brainstorming sessions and much back and forth, Catapult came up with a name they thought checked all the boxes. Relevant parties congregated, a PowerPoint was fired up, and following a dramatic buildup, DustPuppy's new name was revealed: Vroom.

That seemed like a good name. In a simple, single syllable it conveyed dynamism and excitement; it suggested power and speed. But further discussions between Winston, Jeff, and Catapult ensued and over their course a worry developed. Maybe Vroom was more appropriate for the monster truck crowd than the suburban housewives we were hoping to attract (succumbing to the stereotype that those two sets don't intersect).

Diligent Catapult actually developed many candidate names. Among others they pitched after Vroom was Roomba. That immediately caught Winston's and Jeff's attention. Roomba, a combination of the dance "rumba" (also Spanish slang for "party") and "room," seemed to capture the spirit of the robot. It reflected the curious pirouettes and wall-following waggles the robot performed as it cleaned. So, like Norma Jeane rechristened Marilyn, DustPuppy became Roomba.

Any disappointment I felt at the dustbinning of the DustPuppy moniker was short-lived. Roomba sounded OK. And I grew to like it even better over time. In any case, the DustPuppy team now became the Roomba team.[4]

BRUSH UP!

The point of Roomba is to haul a cleaning mechanism around to all parts of the floor. The robot's drive wheels provided the force that moves the robot forward. But since the primary cleaning brush rotated in the opposite direction from the machine's travel, the cleaning mechanism could sometimes hold the robot back. To achieve both proper cleaning and mobility it was critical that this brush press against the floor with neither too little nor too much force.

To regulate that force we needed to adjust the height of the brush relative to the drive wheels—the lower the brush, the more it pressed into the carpet and the stronger the force pushing the robot backward. Set the brush too low and the robot can't move. But position it too high and the robot doesn't clean. The relationship between brush height and force created a problem Roomba—as the robot cleaned, we would need to constantly adjust brush force, and therefore brush height, to accommodate whatever surface Roomba found itself on. Doing this in a frugal way was a major challenge.

Standard practice calls for using sensors to determine how hard the brush is pressing into the carpet and then, through an algorithm running on the microprocessor, command a dedicated motor to raise or lower the brush such that the desired force is maintained. But Dom Pérignon luxuries like fancy sensors and extra motors would bust our Dr. Pepper budget. We needed a much cheaper solution to control our brush.

To the rescue rode Dave Nugent, an iRobot engineer interested enough in the Roomba project to help out on his own time without official sanction. Dave decided to tackle the adjust-a-brush problem.

It felt like there ought to be a cheap solution, as we had a relatively big motor spinning the main brush. When the force on the brush went up, the torque (or turning force) the motor supplied also went up. Dave's breakthrough was understanding that he must somehow use the torque the brush experienced to adjust the height of the brush. He just needed a clever mechanism to transform torque into height.

After a week or two of diligent effort inspiration struck Dave; he constructed and then presented his prototype brush adjusting mechanism to the team. The contraption used a large plastic disk about twice the diameter of the motor, with a spiral groove and a bearing that followed the groove. I remember it being impressively elaborate. On Dave's unpolished prototype, the forces the motor generated weren't quite strong enough to adjust the brush properly, but when Dave moved it with his hands, we could see what the mechanism was intended to do.

We all stared at Dave's mechanism. We tried to follow its logic. We fiddled with it. After a while we concluded: close but no cigar. Dave's torque-adjusts-brush-height concept was sound, exactly what we needed, but his mechanism was too complicated. The concept was fascinating, however, and over the next few days we kept coming back to it. Eventually we arrived at a surprisingly simple alternative.

The flapper and bristle brush were mounted in a carriage, resembling a little box open on the bottom, which could swing up or down. The hinge that enabled the swinging was positioned toward the rear of the robot so the front of the carriage could lift or lower. The main brush motor was also mounted on the carriage, it moved up and down in tandem. The trio of flapper brush, bristle brush, and motor resembled soda cans stacked on their sides.

The output of the motor drove the gearbox that spun the bristle and flapper brushes and, initially, the motor was mounted rigidly to the carriage. Then it occurred to us. What if we changed the motor mounting so that it was free to rotate around its own axis and then wrapped a string around the motor attaching the other end to the robot chassis? With this setup, when the motor encountered greater torque, it would try to wrap the string further around itself. But that would pull the motor and brush carriage upward, reducing the torque on the motor. The carriage would stop climbing when the upward and downward forces matched. Likewise, if the torque on the motor decreased (as would happen if the robot drove from a carpeted to tiled area) the motor would unwind itself, lowering the carriage until the forces again matched. That simple arrangement gave us exactly what we needed. The motor, twisting to a greater or smaller degree against higher or lower loads, would raise or lower itself and the carriage until it achieved the correct force between brushes and floor.

Avoiding both expensive sensor/motor combinations and fragile mechanisms, we managed to reduce the complicated brush-lifting problem to a solution that required one short piece of string. Roomba's brush lifter was one of the more fun and satisfying bits of technology we developed.[5]

ENGINEERING ACROPHOBIA

There's a door in the kitchen of my house that leads down steep stairs into the basement. Although rarely opened, that door was on my mind as the team worked out what we should do about *cliffs*. A cliff was the name we gave to sudden drop-offs like stairs that had the potential to send the robot on an undesired, gravity-assisted journey. To prevent this we would have to give the robot some way to perceive the danger. But a new sensory system would add cost to the robot, and it would give the robot another way to fail. (Eliot liked to say, "What isn't there can't break.") So we asked the same questions we asked of every other system: Can the robot do without it? Is there a less costly way to achieve the same result?

Would it be OK, we wondered, if we made the user responsible for keeping the robot safe? Most of the floor-cleaning robots attempted before Roomba had no way to sense whether they were about to topple over an edge. We were willing to abandon the cliff sensing system if it turned out to be a bank breaker, but we all wanted to make the effort to include that feature in Roomba, especially me with my basement door of doom.

We investigated positioning several sensors on the bottom of the robot, looking downward, spaced just forward of the wheels. As long as these sensors detected a floor, the robot would operate normally. But if one of the sensors witnessed the floor perform a sudden vanishing act, the robot would first slam on the brakes and then spin left or right until the floor reappeared. A reacquired floor would mean that the robot was no longer pointed toward the precipice. The number of sensors required made their cost paramount.

Infrared range sensors made by the Sharp Corporation were commonly used in hobbyist robotics for detecting objects at short distances. High performance and convenient to use, they would make functional floor sensors. Unfortunately, they cost several dollars each; we could afford to spend no more than about a dime per sensor.

For us, however, the Sharp sensors were overkill; they detected an accurate range instead of the binary "floor present/floor absent" we required. We also knew about a sort of a poor man's Sharp sensor that consisted of a LED and a light detector near to each other, aimed in the same direction, and operating in the infrared range. But this kind of sensor could easily be confused by the wildly different reflectances of floor surfaces, e.g., polished white tile versus dark carpet. Confusion could equal a damaging tumble.

Happily, a variation on the poor man's Sharp worked out. Although a ghostbuster might disapprove, we crossed the beams. Or, more precisely, we designed a mounting system that limited the emitter and detector to work in narrow cones and angled them toward each other to make a virtual Venn diagram of infrared light. Only if the floor was located within that limited, intersecting volume would the sensor report that it was safe to proceed. For just a few cents, this trick let us build the cliff safety system Roomba needed.[6]

I used a prototype cliff sensor to do a fun demonstration at a staff meeting. I put a cliff sensor-equipped robot on the table and turned it on. As the robot approached, each person seated around the table stretched out their hands, preparing to catch the robot if it went over the edge. However, Roomba circumnavigated the table, driving right up to the brink, but then always turning away at the last second.

At first, we team members acted similarly to the meeting attendees. While running Roomba on a tabletop back in our office, each of us would tense up as the robot approached an edge, making ready to spring forward and snatch the robot in mid-tumble. But eventually, the cliff-sensing system earned our trust, and we became comfortable letting Roomba operate near edges completely unsupervised.

HELPFUL HERESY

In most engineering firms, standard practice when developing a new product was to begin by writing a *product requirements document*, or PRD. This document spells out, not how a product will be implemented, but rather what, in every respect, the product must do: e.g., the robot shall clean as well as a carpet sweeper; the robot shall clean to the edge of the wall; the robot shall avoid cliffs. The PRD is often regarded as an essential tool for keeping everyone on the same page, communicating between disciplines, and enabling effective project management.

I'd never heard of such a thing. If my teammates were any better informed, they did not let on. Roomba was developed without the benefit (or perhaps tyranny) of such an explicit guiding document. Instead we added and triaged features by consensus at frequent meetings of the team. Because Roomba had no successful precedents there was little relevant history to guide our development. Feeling our way through *terra incognita* we regularly encountered difficult topography. Discoveries included things like the cost of implementing a particular feature, the surprising architectural elements in different parts of the country, and, a discombobulatingly unexpected customer expectation. The delicate and dynamic relationship between what we wanted the robot to do and what we could afford to have it do might have required rewriting the PRD on a daily basis if we'd had one.

In a way we didn't need a PRD because our intimate familiarity with the floor-cleaning domain meant the requirements document was already in our heads. We maintained continuity through repeated meetings and discussions and, importantly, proximity. All of us being in the same small room made it hard not to know what everyone else was doing.

Once I overheard Eliot describing to Phil his idea for repositioning a couple of the cliff sensors so that they could be mounted more conveniently on the bumper. Eliot hadn't considered the change to be significant, but I knew any displacement of the sensors from their current positions would compromise their ability to guard the drive wheels from going over an edge. Our privacy-free office enabled me to call foul immediately, thus avoiding unproductive work. Eliot quickly found another way to mount the sensors, keeping them in their optimal spots.

Those trained in the refined art of product management will no doubt find our PRD-less approach appalling. I've watched a look of horror spread across the faces of such professionals as I explained to them the process we followed while developing Roomba. But for a revolutionary product like Roomba there was a critical upside—we could make far-reaching changes instantly whenever we learned something new about our domain or a constraint on the robot. This happened repeatedly. Described later, the curious incident of the late-project vacuum is the most dramatic example.

Sometimes, some of us (well, me) were almost too willing to change things. I remember not long after we'd made a big update to Roomba that I discovered something new that would make it work even better. Of course, I wanted to redo everything again to include that new learning. Paul became exasperated and said, "With you there's no such thing as a *dirty* slate is there?"[7] That time, I think we held off on making the change I wanted.

Another development sin we committed was our banishment of *modularity*. Modularity effectively puts different parts of a developing product into different silos. This lets subgroups within a team work in parallel without conflict; e.g., the folks developing the drive wheels don't have to worry their design will interfere with the folks working on the cleaning mechanism. All everyone has to do is agree on the boundaries of the silos—how much power each system will draw, what the dimensions are, how one will interface with the others, and so on.

We couldn't do that. Since Roomba had no precedents, we had no reliable way to specify the silos. Systems could easily interfere with each other and often did. Changing something in one system usually required accommodating changes be made in other systems. This was yet another reason we needed constant communication and coordination between teammates.[8]

We did manage a modicum of modularity. From working on the fishing rod toy, Eliot realized that Roomba's intense packaging requirements (everything must fit into a tiny volume) demanded a hybrid approach. The various modules of the system (like the wheel assemblies, cleaning head, and bumper) needed to be independently removable for assembly and debugging, but their intricate outlines were all designed simultaneously to align with each other with *parametric* CAD parameters.[9]

BUMPER

We wanted the robot to clean right up to the edge of every obstacle, touching objects before turning and moving off in a different direction. Sensing objects from a distance using range sensors wouldn't help much with this process.[10] We needed a bumper that could sense contact.

But a bumper was tricky. It has to be big enough to cover the entire front half of the robot to avoid getting hung up on non-sensing areas. It had to be lightweight but rugged because users would sometimes pick up the robot by its bumper. Surprisingly, the bumper had to move not just forward and back but left and right as well in order to detect oblique collisions. And the bumper had to be ultrareliable even though it operated in the presence of dust, dirt, and sometimes pet hair. If the bumper failed, the robot failed.

Paul and I had built a bumper for Scamp with some of these key characteristics. Now we needed something that met all the requirements as well as being compatible with mass production.

Before beginning the production design, Eliot had sensibly picked up the phone and started calling around iRobot. "Has anyone built a bumper that triggers instantly from multiple directions and lasts a million cycles?" he wanted to know. Since he was working at one of the world's premier robotics companies, Eliot was sure that one team or another would have a suitable answer. But all the responses he got were some variation of, "Good luck with that … let us know if you figure it out because we want one too!"

We tried many designs. But with each new prototype it seemed the bumper found a new way to malfunction. For example: The bumper reliably detected obstacles near its center but was blind to obstacles on the sides. Or it was overly sensitive, reporting phantom collisions whenever the robot accelerated or turned. Or it broke too easily when handled. Or it seized up when exposed to dirt. And on and on. After many iterations, we (mostly Eliot) arrived at a design that used linkages to position the bumper robustly, springs to push it into a default position, and optical break-beam sensors that triggered any time the bumper was displaced from its default position.[11]

FIVE-CENT TRANSISTOR

While we were developing Roomba, John, an old friend of mine from college was consulting at iRobot on another project. We encountered each other in the lunchroom one afternoon and began discussing the design. John asked if the robot would include any cool high-tech sensors like phased array sonar or scanning laser rangers. I told him, "We can't afford anything like that on our robot." "But," John said, "don't you wish you could!"

"No!" I exclaimed. "That's the opposite of what we want."

My old friend was expressing a popular sentiment. Most engineers shared John's intuition that more technology would lead to a more refined and better better-performing product. But our team viewed any nonessential technology and its attendant cost as the millstone about the necks of the hapless floor-cleaning robots that flailed and sank before Roomba; we were determined to keep our mind-child metaphorically afloat. This called for hypervigilance—we could add a system, module, or component to the robot only if were strictly necessary, only if the floor could not be cleaned without it. Our mantra became, "Every component must earn its keep." Our team probably spent more time discussing and researching ways to reduce or eliminate cost than we devoted to any other single topic. One example stands out.

Power transistors are the intermediaries between motors and microprocessors. The microprocessor sends a tiny amount of current to the transistor (that's all the microprocessor can manage). The transistor amplifies that signal and sends a large current to the motor (motors take a lot of current to move the robot). One transistor is needed to control one motor. If, that is, all you want to do is turn the motor on and off.

But the robot sometimes needs to spin a motor forward and other times backward. To drive a motor in either direction requires an assemblage of transistors called an *H-bridge*. And an H-bridge must have not two transistors but four. The problem for the team was that the transistors we needed cost five cents, and we wanted the ability to turn Roomba's two-drive motors in either direction. If we only ever spun the wheels in one direction we'd need 10 cents' worth of transistors. But if we insisted on bi-directional control, we would need two H-bridges with four transistors each for a total of eight transistors, or 40 cents.

Forty cents was too much. Forty cents could push us over the threshold we'd chosen and risk making the robot more expensive for customers. So we set about finding a creative alternative. A wild set of thought experiments ensued, including the possibility of splitting up the battery pack into two, making only one drive wheel turn backwards, or other exotic solutions. Nothing seemed workable, yet, grimly determined, we pressed on.

Unexpected rescue came from our representative in East Asia who found a power transistor with the right characteristics that cost a mere one and a half cents. This gave us two H-bridges for only 12 cents. Clouds parted, sunlight burst from the solitary skylight in our office, and we moved on to other challenges.[12]

VIRTUAL WALLS

We expected that users would sometimes want to confine Roomba to one area or exclude it from another, say to focus its cleaning on an especially dirty area or to avoid a cluttered playroom. How should we implement this feature?

At the time there seemed to be only a few reasonable ways to accomplish what we wanted. So I listed them on a whiteboard in the Roomba office for team debate:

1. Confine Roomba with a physical barrier.

2. Mark the floor with paint that only Roomba can see.

3. Define the boundary with a Roomba-detectable magnetic strip.

4. Project a beam of light and make Roomba avoid it.

Each entrant possessed both charms and less appealing features. A barrier relied only on the Roomba's bumper but would need to be heavy (thus increasing shipping costs). Also, anything stretched across the floor constituted a tripping hazard. We did not want to send our customers sprawling. A special clear paint existed that fluoresced when exposed to ultraviolet (UV) light. But we would need to provide the paint, and owners would need to apply it. Eventually it would surely wear off allowing the robot to escape without warning to the user. A magnetic strip would require an adhesive backing to keep it in place, making it a single-use method. Also it would require multiple magnetic field sensors on the robot.

The light beam method could be unobtrusive, easy to position and reposition, it wouldn't cause a tripping hazard, and it wouldn't erode. To implement it we'd need to add just one inexpensive sensor to the robot. But the approach would entail designing another accessory (the device that emits the light beam), the boundary established by one device could only be straight, and the new accessory would require batteries.

The team and I debated, but ultimately the light beam won, and I set about developing a practical implementation. The basic idea was to establish a "virtual wall" that Roomba would turn away from when it was detected. We would distinguish our virtual wall from the rest of the light in the room by using infrared rather than visible light and modulating the light.

Infrared light has a longer wavelength than visible light, so human eyes can't detect it.[13] But wavelength by itself wasn't enough. The sun and other sources (incandescent bulbs for example) produced copious amounts of infrared light. To exclude those (infra)red herrings we would employ the trick of turning our light source on and off at one particular high frequency.[14] Then we would use a detector on Roomba that reacted only to light turning on and off at that same frequency. (We actually used the identical trick to immunize Roomba's cliff sensors and wall follower from ambient distractions.)

We'll call the device that emits the beam the *virtual wall transmitter*. The transmitter relied on an inexpensive IR LED to produce the light, and a very cheap microprocessor (about 11 cents) to turn the light on and off. Now that we can create a beam of light, how do we detect it on the Roomba and turn away from it? One way to avoid the beam from the virtual wall transmitter would be to add several detectors distributed around Roomba's sides, arranged in such a way that when any detector saw the beam it commanded the robot to begin spinning in place. The robot would continue spinning until no detector saw the beam. Then the robot could drive forward again, certain that it would not cross the beam.

I fiddled with different configurations, but they always ended up using too many detectors. More than a couple of detectors would make the electronics too expensive. It turned out, however, that there was an arrangement that required just one sensor. If I put an omni-directional sensor on the top front of the robot I could achieve my thrifty ideal.

Since Roomba only drove forward, the nose would always encounter the beam before the center of the robot crossed the beam, even if the robot encountered the beam obliquely. As soon as this detector saw the beam, the robot began to spin in place until the receiver exited the beam. It was then safe for the robot to proceed.[15]

That solved most of the problem, but a couple of details added complications. The beam emerged as a ray[16] from the virtual wall transmitter. The robot could see the beam only when it crossed the beam. But what if the robot approached the transmitter from behind, or from the side? It was possible for the robot to bump the virtual wall transmitter without seeing the beam. If that happened, it might displace or twist the transmitter and leave the beam pointing in the wrong direction. To fix this, we had to add one more feature to the virtual wall transmitter—a low power omni-directional IR emitter modulated at the same frequency as the beam. With this, anytime the robot got close to the virtual wall transmitter, it detected IR and turned away before it bumped the transmitter.[17]

THE BIG LIGHT

Colin sometimes dropped by the Roomba office to check on our progress and talk about the robot. On one such occasion the discussion turned to spot mode—the behavior intended to clean an isolated patch of dirt. Colin thought it would be a cool idea to install a large, bright LED on the underside of the robot pointing straight down. This, he imagined, would help users position the robot to clean the spot.

The team was dubious. As we were clawing and scratching over every penny of cost going into the robot, we didn't think a big light would add enough value to justify the expense and design effort. But we'd just heard from the CEO, so we installed a fat, bright LED on the bottom of a prototype robot and tried it out. The light shining on the floor indeed looked cool. But we remained of the opinion that the feature didn't add enough value to justify the cost and effort.

HIGH CENTER

Any small robot or vehicle that operates close to the ground runs the risk of *high centering*. If the robot approached a shallow, narrow wedge (like one foot of a spider-leg table), momentum would drive the robot up the wedge lifting the front. If the drive wheels are rigidly attached to the chassis, the robot could easily find itself partly suspended with its wheels spinning helplessly.

And now it's stuck. Even if onboard sensors inform the robot that it has stopped making forward progress, there's nothing the robot can do.

Reversing the wheels won't back it down from the wedge; the robot's wheels just spin in the air. This situation is called high centering.

Spider-leg tables are relatively rare, but wedge analogs are common. A child's flip-flop, a small toy ball, a pile of papers stacked in an unfavorable way can all serve to ensnare the robot. Any feature like these can become jammed under the robot in such a way that the robot's wheels are prevented from making firm contact with the ground.

To improve the robot's ability to negotiate real world (or real home) terrain we needed some kind of *suspension*. We couldn't just attach the robot's drive wheels rigidly to the chassis.

The simplest, cheapest solution we could devise was to build a module that included the drive wheel, motor, and gear box, attached to the robot so that it could pivot down.[18] A spring kept tension on the module. The size and stretch of the spring were adjusted so that it applied *almost* enough force to counter the entire weight of the robot and lift it up. Now if the robot ran over something that threatened to high center it, the drive wheels would not immediately lose all traction. Instead the wheel module would swing down, keeping the wheels in contact with the floor.

To which side of the wheel module ought we attach the pivot, front or back? We built a prototype with the pivots toward the back, behind the wheel modules. Then Paul and I conducted a bunch of tests—forcing the robot to drive over all sorts of difficult objects and into challenging situations. We tested the mechanism commanding the robot to drive both forward and backward. Thus we tested the wheel modules with the pivot both leading and trailing.

With the wheel module pivots in the back, the robot could drive over nastier collections of objects going backward than it could traveling forward. I considered asking Eliot to change our initial design, switching the position of the pivots from the back to the front to give Roomba better forward mobility. But then I thought better of it. It made sense that the robot should be more adept at backing out of trouble than driving into it. Giving the robot greater ability to extract itself from a difficult situation and less to get into one should mean Roomba would become stuck less often.

SARA FARRAGHER

Sara Farragher liked art. She liked drawing, painting, and crafting. Creative activities brought her such joy that even before high school she'd already concluded, "I will become an artist." But the rigors of art school and the realities of the world after graduating from RISD[19] revealed a hard truth. Making art and making a living from art were entirely different, even contradictory undertakings.

Sara landed in an administrative job at Harvard, but she soon began to feel that the fit was awkward. Once, she and her co-workers were given the Myers–Briggs personality test. Afterwards, the test proctors posted a plot of the results. In one quadrant of the personality map was a tight and convivial constellation of dots. At some remove, in a different quadrant, was one lonely little dot. Sara's dot was the lonely one.

She decided to try something else.

Through the almost-new internet, in the form of Monster.com, Sara discovered an open position at iRobot and spoke with Winston. Their conversation went well, and Winston offered her the job. Half of her time, he explained, would be spent on prosaic things like processing purchase orders and invoices for the Roomba team, but the other half would involve more creative pursuits, often working with marketers. The creative parts of the job were especially enticing, so Sara accepted the offer.

Winston wanted Sara and the team to meet, but before that introduction Winston spoke with us. He issued a stern caution telling us that Sara was young, and we should not come on too strong. "Don't frighten her away!" he said. But the initial meeting was great fun, and we all ended up sitting on the floor looking at pictures in Sara's art portfolio. Far from being put off by the unrefined behavior of the barely civilized engineering team, Sara found the atmosphere in the Roomba office reminiscent of pleasant childhood episodes. It called to mind hanging out with her father or older brother while they tinkered with their projects.

Given the choice of a desk near the iRobot finance folks or being shoe-horned into a cubicle in the cramped Roomba office, Sara picked the latter—to the benefit of all. Sara's presence soon proved, if not the glue that bound the team together, then the WD-40 that smoothed all interactions. The Roomba office registered a measurable increase in laughter and joy after Sara arrived.

***

Energy and excitement permeated the overstuffed closet that was the Roomba office. Even our visitors sensed it. Constantly there were new ideas to discuss and test, new challenges to meet, new mechanisms to invent. Jokes and personal stories[20] peppered discussions as we worked. Playfulness ruled: Chris named one of our prototypes the Robot of Eternal Happiness, I mounted a giant cutout of Bullwinkle J. Moose[21] above my desk.

Conversations ranged over subjects far and wide—Eliot explained why trains don't have clutches, Paul taught us to count music in weird 7/4 time. Rather than work, it felt like hanging out in a cool clubhouse. Roomba was fun. Even aside from our historic quest, the sheer spirit of the place made me eager to come to work each day.

That happy vibe confirmed the demise of one foe. Whether through design or providence, we had exorcised the wretched wraith of Clean. And although we *did* argue—often and passionately—disagreements never tested our unity. Evidence was the arbiter of every debate. We remained of one purpose. We all felt it; *we can do this thing!* Even Chris's pessimism began to waver.

NOTES

1 To date, I know of no significantly shorter robot vacuum cleaner that has reached market.

2 A computer aided design program lets the user build mechanical parts virtually on a computer. For his work Eliot used a program called Pro/Engineer running on a workstation optimized for CAD.

3 We gave the Dynacept Corporation huge amounts of business. Dynacept reportedly considered naming one of their machines after Eliot. His late-Friday phone calls followed by an avalanche of uploaded parts became legendary at the company.

4 A story repeated occasionally in the press claims that Roomba was very nearly named "Cybersuck." It was not. Sometime during the first few years of IS Robotics we conducted an informal exercise trying to think up robots we might build and what we might name them. A vacuuming robot was an obvious candidate and someone (maybe me) farcically suggested Cybersuck as a name. A roguish coworker thought that moniker would also be appropriate for an entirely different line of (shall we say prurient?) robotic products that he advocated. The joke may have resurfaced while we were working on Roomba, but the name Cybersuck was never a serious contender.

5 The brush-lifting mechanism is disclosed in patent US7571511B2 For simplicity, the text above describes the string as wrapping around the motor (it did in our initial concept) but in production the string wrapped around a spindle in line with and rigidly attached to the motor. To adjust the torque to a desired value we needed to control the diameter of the item the string wrapped around; ergo, the spindle.

6 Patent US8788092B2.

7 Paul was reacting to my constant desire to start with a clean slate.

8 Our omission of the PRD and modularity may have served Roomba well, but sadly, such heresy doesn't scale. A development that requires a big team needs formal structures that a small team can implement informally. Or such is the conventional wisdom.

9 Using parametric variables lets the CAD program do some of the work for the designer. Adjusting the dimensions or position of one component can change other components automatedly. But a manual sanity check following a change is always required.

10 The only utility a range sensor might provide is to alert the robot to slow down just before it bumps into an object, thus reducing the force of collision. Curiously, Paul and I built that ability into Scamp. We used very simple infrared sensors to detect the reflection from a nearby obstacle. These sensors were good at detecting walls but often missed small things like spindly chair legs. Because they only worked some of the time and because of their cost and complication, we eliminated these rudimentary range sensors from the first iteration of Roomba. Years later iRobot restored this feature in some models.

11 The bumper is disclosed in the same patent as the brush lifter, US7571511B2.

12 We might have been a little less adamant about the cost of transistors had the issue come up later in development. Early on we were trying to keep the robot's retail price under $100.

13 Astronomer William Hershel discovered infrared light in 1800. He used a prism to split sunlight into the colors of the rainbow and then systematically positioned a thermometer to intercept each color in turn. As he moved from blue to red, he noted that the temperature kept increasing. But astonishingly, the highest reading of all came when he placed the thermometer in the dark just past red. The thermometer detected light no one knew existed, that neither Hershel nor anyone else could see. Discovering that must have been so cool!

14 Roomba looked for infrared light that turned on and off 40,000 times per second.

15 Patent US8368339B2.

16 The ray, in the zero-diameter mathematical sense, was too strong. The beam needed to be fat enough that Roomba couldn't drive all the way through the beam before it had time to react.

17 I was both proud of and embarrassed by the robot confinement solution we settled on. The sensor on the robot used the phenomenon of total internal reflection and a clever rotated parabola shape to direct photons to the detector—which my inner physicist loved. And it needed only one detector— my inner miser loved that. But the required topknot made the robot almost 4 mm taller than it otherwise would have been. That this reduced mobility somewhat has always grated on me.

18 Roomba's suspension system was somewhat like that of a car. A car's wheels can move up (to absorb bumps) and down (to maintain contact with the ground). Our robot's wheels needed only to move down.

19 The Rhode Island School of Design (RISD) is a prestigious art school in Providence, RI.

20 I am forbidden from telling the one about the frying pan and the sibling, likewise the one about playing chicken with safety barrels on the interstate.

21 *The Adventures of Rocky and Bullwinkle and Friends* was my favorite cartoon show growing up.

CHAPTER 13

Reality Bites

As Roomba prototypes became more operational, we were able to expose the robot to more challenges—to customers who might buy it, to homes and floors where it would ultimately perform its service. These in turn exposed many flaws in our thinking.

NATURE ABHORS A CARPET SWEEPER

One pleasant mid-summer day Roomba's engineers, Winston, and several other iRobot folk rendezvoused at an unremarkable, multi-story office building on the Cambridge side of the Charles River. We assembled in a narrow room. A long table occupied the room's center. Snacks and sodas were set out along the back wall; the lighting was subdued. The dominant feature of this cramped chamber was a big one-way mirror occupying almost the entire front wall. Sitting at the table, one could see through the looking glass into a wonderland of market research on the other side. In that much larger, brightly lit room were comfortable chairs, an easel with a large pad of paper, and our hired facilitator. Although this was a familiar trope I'd seen a hundred times on TV, actually lurking in an observation room like this felt a touch surreal.

We'd paid maybe $10,000 for the privilege of setting up a focus group—probably the most the company had ever spent on a market research event. But we needed to know how potential customers would react to Roomba when they saw one in the (plastic) flesh cleaning the floor at their feet. At the appointed hour, our facilitator welcomed eight to ten bona fide ordinary people as they filed into the large room and sat in the chairs. Our mind-child was about to receive its first critical judgment from strangers.

 DOI: 10.1201/9781003540489-13

The facilitator prepared participants by encouraging them to state their honest views and not to be swayed by the comments of others. "You are the world's expert in your own opinion," she told them.

At first the facilitator described Roomba without showing the group any photos or the device itself. She was met with skepticism that such a thing would actually work. Then she demonstrated one of the prototypes we had prepared for the event. As participants watched Roomba go about its business their doubts ebbed. Even those who stated that they would never purchase such a device couldn't help being intrigued. As the group discussion proceeded, soccer moms emerged as the most interested. They saw Roomba as a time saver. This surprised and pleased us, as we'd expected the much smaller market of gadget geeks would be the first to fall in love.

But we could take neither interest nor love to the bank. We needed to know how much customers would pay. Our facilitator eased into that part of the gathering's proceedings. She did not inquire directly, but rather asked, "If you saw this product in a store, what would you expect the price to be?"

The focus group's responses were all over the map. Some people mentioned a price close to the $200 we intended to charge. A few folks we regarded as saints-in-training expected an even higher number. But most were lower. One woman said she'd expect Roomba to be priced at $25. Later when asked what she thought a replacement battery might cost she said, "$50." That hurt. For this lady, attaching our robot to a battery devalued the battery.[1]

Throughout the proceedings our facilitator had been careful to leave a couple of things unmentioned. First, she never referred to Roomba as a robot, calling it instead an "automatic floor cleaner." Three separate groups, comprising an aggregate of around two dozen people, gave their opinions that day. Of these, only two individuals spontaneously applied the term "robot" to Roomba.

The second unmentioned characteristic was the nature of Roomba's cleaning mechanism. That is, the facilitator had revealed no details about how it worked. Participants had seen the demo, they observed Roomba cleaning effectively, they had given their opinion about the price. They'd all assumed that a vacuum was at work, several used that term to refer to the robot. But now the facilitator told them, "Roomba is a carpet sweeper not a vacuum." Then she asked again what they would expect to pay.

On average, focus group members from all three groups cut their estimates in half. Participants who had previously said $200 now said $100.

Kaboom!

The focus group's brutal revaluation exploded our world. The enabling innovation that made the energy budget work, that made Roomba technically and economically feasible, was cleaning with a carpet sweeper rather than a vacuum. People had seen that the carpet-sweeper-Roomba really did work. Yet, they chose to trust conventional wisdom rather than their own apparently lying eyes. If we were forced to cut the robot's price in half, we would lose money on every unit sold, and there would be no Roomba.

At the end of the evening before any member of our stunned team could stagger out the door Winston said simply, "Roomba has to have a vacuum." A shotgun wedding was in the offing for bot and vac.[2]

The next day at work we gathered to discuss the focus group's revelation. A halfhearted attempt or two to deny reality quickly faded, Chris Casey saw to that, and we accepted what we needed to do. But changing things now would be a huge challenge in multiple ways. We were deep into development, closer to launch than kickoff. All the electrical power our battery could supply is already spoken for. None was available for a new system that would likely be more power-hungry than all the robot's other systems combined. And where could we put a vacuum? All the space in the robot was also fully assigned. Our mandate to clean under furniture and between chair legs wouldn't let us make the robot any bigger.

One escape hatch beckoned, but no one was eager to leap through it. Chris articulated what we were all thinking. "We could build a vestigial vacuum," he said. That is, we could design a tiny, pico-power vacuum—one that consumes almost no power and does almost nothing—strap it on the robot, and call it done. Perversely, that seemed reasonable. The robot already cleaned the floor well; our cleaning tests prove it. Customers, however, didn't know that. They were all steeped in the dogma of vacuum supremacy. Re-educating the masses wasn't possible—we didn't have the funds. But if we could assert on the box that Roomba had a vacuum, then everyone would be satisfied. We could charge the price that makes our economics work. Customers would deem that cost reasonable and wouldn't have to unlearn their vacuum bias.

But it felt wrong. If we must add a new system to the robot we wanted it—like all the other systems—to earn its keep honestly, to do something useful. Through further discussion and calculation, we concluded that we could afford to devote about 10 percent of the robot's 30-watt power budget to a vacuum. Conventional manual vacuums typically gorged themselves on 1200 watts of power, but if we could develop a system that provided useful cleaning while consuming only three watts (0.25 percent of 1200) then we would feel good about adding it to the robot. It just didn't seem very likely.

I sometimes find that solving a problem is largely a matter of staring at the problem's source. Gaze long and intently enough at something and, Waldo-like, the solution may reveal itself. So I took one of the team's manual vacuums and stared at it. What exactly made it use as much power as it did? I knew the answer was partly marketing rather than reality. There was no simple, objective way to compare cleaning efficacy between vacuums. Lacking a results-based method, shoppers looked at inputs. For example, a vacuum with a 10-amp motor sounds as though it should clean better than a vacuum with a 6-amp motor. But the bigger number might only mean that the manufacturer with the 10-amp claim was using a less efficient motor—the 6-amp (720 watt) motor might clean just as well.

But even when you corrected for the amperage arms race, a vacuum was still a power glutton. Staring at the vacuum cleaner, I began to see why. The vacuum fixed in my gaze that day used the standard configuration: a cylindrical beater brush occupied the center of a wide air inlet. A motor, attached by a belt, spun the brush. Another motor, deeper in the machine, drove a *centrifugal blower* that drew air in through the inlet. To keep dirt particles kicked up by the beater brush entrained in the airstream, the air needed to move fast. The combination of a wide inlet and high velocity meant that every second the vacuum motor had to gulp a huge volume of air.

Accelerating all that air took considerable power—the physics was inescapable. If we wanted a vacuum that sipped power rather than guzzled it, we had to move a much smaller volume of air per second. We could accomplish that—without reducing air velocity—if, instead of a wide inlet, we used a narrow one. To match the manual vacuum's air velocity using only a three-watt motor, I computed that we would need a narrow opening indeed: only a millimeter or two.

That instantly disqualified Roomba from using the standard vacuum configuration—we could not put our bristle brush in the middle of the air inlet. That would require an inlet maybe 20 times too wide. We'd have to find another arrangement.

To test the narrow-inlet idea I turned to my favorite prototyping materials: cardboard and packing tape. Using these, I mocked up my idea.

The inlet for my test vacuum was as long as Roomba's brush but only about two millimeters wide. To provide suction I repurposed the blower from a defunct heat gun. Then I applied my jury-rigged contraption to crushed Cheerios and a variety of other dirt stand-ins. My novel vacuum was surprisingly effective at picking up small debris from a hard surface. Using an *anemometer* to measure the speed of the air rushing through my narrow inlet showed that it was, as desired, as fast as the air stream in a standard vacuum cleaner.

The next step was to somehow shoehorn our micro-vacuum into Roomba. To form the narrow inlet we used two parallel vanes of rubber. Small rubber bumps protruding from one vane spanned the inlet preventing the vanes from collapsing together when vacuum was applied. We placed the air inlet parallel to and just behind the brush. The only plausible space for the vacuum impeller,[3] motor, and filter (needed to separate the dirt from the flowing air) was to take over a corner of the dust cup. Drawing on his now well-honed skills of packing big things into tiny spaces where they had no business fitting, Eliot managed somehow to accomplish this. But we did get help from an outside consultant to design the intricate shape the impeller needed to move air efficiently.

In general, regular vacuums perform better on carpet than on hard floors. But Roomba inverted that relationship. Our vacuum operated like a squeegee, pulling dirt from tile, linoleum, and wooden floors. But it was less effective on other surfaces. The sweeper mechanism did the heavy lifting when cleaning carpet.

Despite the team's reluctance to add a vacuum and despite the unit's low power, the vacuum genuinely improved Roomba's cleaning ability. We could demonstrate this convincingly. First, we disabled Roomba's new vacuum by disconnecting the power and then cleaned a hard floor relying only on the carpet sweeper mechanism. If we then walked across the floor barefoot, we would feel a certain amount of grit under foot. If we repeated the exercise with vacuum power on, the floor was pristine. Bare feet would detect no grit whatsoever.

Years later I learned that the focus group had a back story no one mentioned at the time. While the Roomba team had swallowed the carpet sweeper concept hook, line, and sinker, Winston had not. He was uneasy with the notion that customers would be cleaning-mechanism agnostic—thinking instead that they simply wouldn't believe our robot would clean their floors if it didn't have a vacuum. He found at least indirect support for that position when he scoured marketing data SCJ had shared.

But Winston, well-attuned to the engineering psyche, knew he couldn't just declare, "Roomba has to have a vacuum." We'd have pushed back, probably saying something like, "What your business school-addled brain doesn't appreciate is that it's the carpet sweeper that makes the whole concept work!" Winston, managing a team of one genuine and several virtual Missourians, had to show us. That was a key purpose of the focus group, to demonstrate to the Roomba team that we had made a deal-breaking omission.

Touché!

FMEA IS A FOUR-LETTER WORD

FMEA, short for Failure Mode and Effects Analysis, is a systematic process engineers sometimes apply to a product under development. During the exercise, participants examine every system and component of their product and ask, "In what ways might this part fail, and what would happen if it did?" The procedure is beneficial because it can uncover risks before they blossom into failures—it thus enables low-cost mitigation. The soul-crushing tedium it imposes on those involved is the process's only disadvantage.

SCJ insisted that we subject Roomba to an FMEA. The exercise was carried out over several weeks and involved meetings in both Somerville and Racine. The meetings were not popular with our team, but whether in Somerville or an airplane ride away in Racine, patient-as-a-saint Phil attended every one. At the one meeting Chris was unable to avoid, he grew to loathe the phrase his SCJ inquisitor repeatedly used: "Let's capture that." He decided a more accurate acronym for FMEA might be Futile Meaningless Eternal Agony.

While he was still working with the Roomba team, Dave Nugent was roped into one of the FMEA sessions. There Dave displayed a normally admirable characteristic common to many conscientious engineers: honesty to a fault.

In earlier team discussions back in the office we had tried to imagine nightmare scenarios. One involved fireplaces. Most houses with fireplaces have a hearth that is elevated relative to the floor. If Roomba were to clean the floor in such a house, it would bump into the raised hearth and then turn away. But in a much smaller number of mostly older homes the hearth is flush with the floor. In these residences, if neither screens nor closeable doors blocked the firebox, then Roomba would be free to drive right in. The nightmare was that the robot might do so while a cozy fire was burning unattended. Roomba might then collect embers, drive out of the fireplace, and scatter the burning coals around the room.

Much to the alarm of the other team members present, Dave began to describe the catastrophic details of this remote possibility. Subvocalizing, Phil advised, "Dave, shut up!" But Dave went on. As he spoke, the SCJ facilitators nodded solemnly and took careful notes.

After the meeting, we worried that Dave's candor might prompt SCJ to add new expensive requirements to the robot. But none were forthcoming, we dodged that bullet. And Roomba, to our knowledge, has never lived the nightmare we conjured.

GOODBYE, RACINE

Despite their insistence that Roomba incorporate a consumable, SCJ was generally very helpful. The representatives we dealt with were always on-call, always willing to help out with anything we needed, and they gave us access to customer surveys and mountains of invaluable marketing data SCJ had collected in the course of their consumer products business. Winston especially found it a pleasure to work with them.

But then things changed. A reorganization at Johnson headquarters put different people in charge of the project—people who were not convinced that Roomba was a good project or even that SCJ should be in the robot business at all. They measured Roomba not as a revolutionary development that might open doors to new markets but as if iRobot were a vendor supplying some chemical used in the manufacture of one of SCJ's existing products. In that light, Roomba didn't look good.

So, for the second time, SCJ pulled the plug on a project with iRobot. It was a moment of some trepidation for the team—would the project continue to be funded? We all knew money was tight.

Fortunately, for some time Helen had been working relentlessly behind the scenes to secure investment from venture capitalists. A Herculean (Hellenean?) effort was required but she ultimately wore them down. So, by the time of SCJ's second departure from iRobot, enough money was in the bank to make one attempt to complete our project. It was not Roomba that attracted the cash; VCs maintained their skepticism of robot vacuums. But through Helen's perseverance they became convinced that iRobot could find some path to success.

Our latest divorce from SJC gave us custody of the accursed consumable dust cup. Like peasants overthrowing a despot, we exiled it immediately and permanently. Roomba would go forward with the reusable dust chamber the team had always intended to rely on.

WORST DAY

On a crisp, pre-autumn Tuesday I was engaged in my usual weekday morning routine, steering my Toyota Corolla along Massachusetts Route Two eastward toward Somerville. I had the radio playing and a brief bit of news came on saying that an airplane had crashed into a skyscraper in New York City. Probably a disoriented Cessna pilot, I guessed.

But it wasn't a Cessna. By the time I got to work on September 11, 2001, the dimensions of the unfolding tragedy were becoming appallingly clear. The mood at the office was subdued and somber. Some of us were glued to the TV. I sat in my cubicle defensively avoiding the TV.

At that time, the company was developing a very tough and capable, mostly teleoperated mobile robot called PackBot, invented by longtime iRobot mechanical engineer Chi Won.[4] Designed for use by police in an urban environment or by soldiers on a battlefield, PackBot was rectangular and flat, about two feet long by 20 inches wide and 6 inches tall. It was propelled by bulldozer tracks spanning the length of its left and right sides and it included two similarly tracked leading flippers. The flippers could pivot up or down to help the robot climb or descend stairs or traverse torturous terrain. PackBot was nimble, it had no trouble climbing grades of up to 173 prcent (60 degrees), steeper that the steepest ski slope. It could cross railroad tracks, drive inside large sewer pipes, and conquer rock piles. We regularly demonstrated the robot's ruggedness by tossing it out of a second-story window. It just kept going.

Lieutenant Colonel John Blitch of DARPA,[5] one of PackBot's sponsors, had been in touch with authorities at Ground Zero in New York City. Around mid-day he called to ask iRobot for our assistance. To aid with the rescue/recovery efforts he wanted us to send as many robots as possible, along with people to operate them, down to New York. LTC Blitch told us they had to be there that evening.

Everyone at the company was more than eager for a chance to help. We had a couple of robots suitable for deployment, but they needed to be outfitted and configured for the difficult task ahead. The robots would likely be operating in tight, collapsed areas amid abundant dust. This made us worry that dust would settle on the camera lens, eventually blinding the operator guiding the robot. So I, consulting with other folks, began working on a mechanism that would blow or brush dust from the lens.

PackBot communicated wirelessly in open areas but if it drove underground or within a rubble pile it might find its control signal blocked or unreliable. To make sure of continuous communication we would connect the robot to its control console via a long cable. But there was no time to build a proper mechanism to spool out and retrieve the cable. The quickest/simplest solution was to coil the cable into a container on the robot's top and have it pay out as the robot drove forward. Because time in the field would be critical, a cable could be used for only one out-and-back mission. Afterwards the tangled cable would simply be left behind—the robot could not rewind the cable, nor could we burden the operator with retrieving it. The number of cable assemblies we supplied would equal the number of times the robot could be used, so we wanted to prepare as many assemblies as possible.

It turned out that the most appropriate cable-container we could quickly identify was a large round cookie tin—big enough to hold plenty of cable and squat enough to sit comfortably atop the robot. iRobot folks were dispatched to buy up as many appropriately sized cookie tins (and the incidental cookies) as were available in eastern Massachusetts and southern New Hampshire.

Paul, through his work in the theater and boat yards, was well acquainted with how to coil the cable so that it would not become kinked as it paid out from the moving robot. He and several fellow engineers prepared ethernet cables and wound them into the hastily acquired cookie tins.

Everyone else worked on aspects of the deployment that could benefit from their special talents. The level of activity at the company that day was more intense and focused than I'd ever experienced before. No time was wasted, the simplest and most straightforward solutions to all the problems we anticipated were implemented expeditiously. Collaboration ruled the day. Within just a few hours we completed our complex assignment. Engineers then loaded the robots into a van and began the drive to New York.

Ultimately, our efforts were of less direct help than we'd hoped. The robots were intended to assist by exploring voids and interstitial spaces within the rubble. But the destruction caused by the falling towers was so complete that very few such places existed. There were only scant situations where we could put our robots to work.

Most people across the country on that awful day had nothing to do but worry, grieve, and endure. But we at iRobot received a precious gift. We got to *do* something. Working to prepare the robots for their mission allowed us to take immediate action, applying ourselves and our talents to an effort that we knew might help others. If nothing else, it helped us.

And there's a little more to the story. Not everyone was in the office that day. Winston, Colin, and Glen Weinstein, iRobot's chief council, were in Racine, Wisconsin, 1000 miles from home, negotiating SC Johnson's withdrawal from the Roomba project. As news from New York filtered in throughout the morning discussions slowed down.

Eventually, the home team trio decided their most prudent course would be to return to iRobot headquarters. But that was challenging—all air traffic in the country had been shut down. And, of course, many travelers who now couldn't fly were all trying to rent cars and secure other transportation at the same time. It was in that moment that Sara proved her mettle and her worth. Relentlessly working the phones, she managed to find a car for our remote negotiators. A Ford Crown Victoria (stereotypically your grandfather's car) would carry them home.

Taking shifts behind the wheel as they drove through the night, Winston, Colin, and Glen found the roads mostly deserted. As dawn broke the following morning, they were eastbound on the Massachusetts Turnpike, nearing home. It was then that they saw the only other car on the road for miles around, a state police cruiser approaching at high speed on the westbound travel lanes. After the vehicles passed, the police car turned off the pavement, darted across the grassy median of the divided highway, and rapidly closed on the travelers. With sirens blaring and lights flashing, it overtook the Ford, forcing it to stop.

Police officers surrounded the Crown Vic, front and back. There was much shouting. Several tense minutes passed while hands rested on holsters and pointed questions were asked. But once IDs had been checked, the travelers, to their great relief, were allowed to continue their journey home. Did young Winston, Colin, and Glen attract police attention that day because they were simply too hip to be driving an old man's car? Official documentation is incomplete.

REAL-WORLD TESTING

An airplane can be designed and its flight simulated using computer software. The equations[6] that govern the interaction of the wings, stabilizer, rudder, and airframe with the air are well understood. A powerful computer can easily solve those equations. Before it ever takes to the skies, aeronautical engineers can accurately predict how the aircraft will behave, and they will know its key operating parameters.

Robots aren't like that. No equation adequately describes their complex interaction with the world—wheels that slip sometimes, carpet nap that compresses shifting the robot sideways, an item of clothing that—heaped the right way—high-centers the robot. Advances have been made in recent years, but Rod Brooks supplied the adage we followed when Roomba was developed. Of robots he said, "Simulations are doomed to succeed."

What Rod meant was that you include in a simulation only those things you think are important. Then you build your robot's computer model so that it will handle just those important things. In simulation it all works perfectly! The simulated robot solves the simulated problems you created for it. But the real world refused easy pigeonholing. Like an unbottled genie granting a wish, there was always a twist; a subtle, incorrect assumption you made that the real world, snickering to itself, relishes exploiting.

A real airplane designed according to the results of a simulation will soar the first time. A similarly designed robot will not follow suit. Robot simulations can offer hints, but they never provide the last word. For that you must test.

We wanted Roomba to work in every venue. To make sure it would, we tested. Testing started in the lab where we tried to replicate the "typical" home. We assembled tile and carpets of various kinds, populated the robot's work area with chairs and other furniture, and we engineered steps and cliffs. All of us spent hours watching Roomba navigate our test floors.

An iRobot intern, Dan Roth, built a camera/software system to record Roomba's motions. For this we used a big room in the basement of our building—near the place where Phil had tamed the misbehaving Clean robot. Dan's system used a computer to process the video so that it showed everywhere Roomba visited during its run. This let us check for problems of systematic neglect.

We set up life-test pens. Take a four by eight-foot plywood sheet, cover it half with carpet, half with linoleum, and then nail a short wall around the perimeter to prevent Roomba from escaping. Build scaffolding so several such pens can be stacked on top of each other. Modify the robot to accept power from a tether dangling above the pen. Then just let the robots run. We kept them running until they broke or wore out. When either happened, we would figure out what component had caused the failure and then redesign it in a beefier way.

We quickly reached the point where Roomba navigated our test areas flawlessly. It could run for hours and never get stuck. But I knew from my long association with robots that this happy situation was too good to be true. Roomba's in-house success was just an artifact of our limited imagination. We were testing the robot only in the ways we had thought to test it, but reality was so much richer than what we could envision. Exposing Roomba to a wider world would be the only way to reveal the many things we'd missed. Our first outside-the-office testing venues were the homes of iRobot employees.

Jeff Ostaszewski, Roomba's marketing specialist, was a recent college graduate. But as often happens in that transition, early in his professional career he continued his student lifestyle. Some of our team decided to test Roomba in the low-rent apartment Jeff shared with several roommates a 10-minute walk from iRobot.

I remember the floor of the common area in Jeff's flat. It featured a once-proud carpet that now, long into its dotage, had become misshapen and threadbare, much more in need of disposal than cleaning. The carpet achieved only limited success concealing its underlayment—a cement floor that, filled with cracks and potholes, resembled Boston roads at the end of a tough winter. Despite its stellar performance in our in-office test areas, here Roomba became hopelessly stuck within minutes. After our first visit, Paul started calling Jeff's apartment "The Crack House."

It was a term of endearment—we loved Jeff's place! It gave us exactly what we needed, an extreme, real-world test for Roomba's mobility and cleaning systems. It took days or maybe weeks, but we improved Roomba to the point that it did much better on our next visit.

We searched for more torture tests and found two close to home. Visiting Chris Casey's abode we discovered, in the second-floor bedroom, a carpet that was more than a match for Roomba. When the robot tried to drive straight across it would make only painfully slow progress and always arced to one side.

An even more perplexing carpet lurked in the hallway of Winston's Cambridge apartment. To begin his testing regimen, Winston took home a Roomba and set it to run on his 40-year-old, innocuous-looking, beige, medium-pile carpet. The robot began moving sideways. Amazed by the unexpected phenomenon, Winston excitedly reported his results to the team the following day.

We informed Winston that he had made a mistake. Using equations and vector diagrams, we patiently explained (to the man with a PhD in physics) that basic physics disallowed such motion.

Stung by our doubts, Winston went home that night and used a utility knife to cut out a large section of his discord-promoting carpet. He brought it to the office the next day, gathered the team, and set Roomba to work. The robot scooted forward a bit, bogged down, and then juddered to the left.

"It's moving sideways … How is that POSSIBLE?" Chris pleaded. For Winston, vindication was never sweeter (well worth the $100 he spent replacing his landlord's carpet). We subsequently spent many hours tweaking Roomba to enable it to run on a carpet as manifestly evil as Winston's.[7]

Physics-defying carpets were only one of the mobility issues revealed by home testing. Another arose from an unexpected architectural feature. Winston took a Roomba along on a Christmas visit to his sister's brand-new home in Chicago. On its first running, Roomba spiraled out from its starting point, took off across the floor, and encountered the wall at an oblique angle. There it stopped.

The sharp corner of Roomba's bumper had become ensnared on the *quarter-round* molding that lined the walls of every room. We'd never seen such a thing in New England, but in the Land of Lincoln this apparently common design element frustrated Roomba's ability to follow walls. We learned this relatively late in development and the only quick fix was to take a bite out of Roomba's bumper. The gracious curve that marks the left and right sides of Roomba's bumper was put there only to accommodate Chicago quarter round.

MEANWHILE, BACK IN VÄSTERVIK

Trilobite's makeover and its appearance on *Tomorrow's World* turned the tide for the Electrolux robot. Where the Floor-care department had once doubted the robot, they now embraced it and applauded Per Ljunggren's remarkable triumph. Orphan no more, Trilobite joined the production schedule. This was a bittersweet moment for Per. Sweet because the battle he had waged for so long was now decisively won. But bitter because the project was no longer his.

The next step was to take the design Per's team had created and reengineer it to facilitate mass production. An entirely different set of engineers assumed this responsibility. The "industrialization" team was based not at headquarters in Stockholm but in the factory town of Västervik, a 175-mile drive to the south.

The conversion would prove challenging. Trilobite was a complex machine with many precise requirements. Also, Electrolux's standards were exacting. When Trilobite took its place in the pantheon of Electrolux products, it must appear no less godlike than its siblings. In every way, Trilobite must match the reliability and quality that consumers expected of Electrolux. The factory in Västervik possessed all the sophisticated capabilities that might be needed to ensure this. For example, to build their products, most manufacturers simply buy standard motors from a motor vendor. But, if need be, the Västervik factory could construct, from scratch, a motor of any specification.

The advanced nature of the robot was one reason Electrolux chose to manufacture Trilobite domestically rather than move it to a low-cost producer in the East Asia. The second was their worry that their robot would be copied. Patents and copyrights are respected in Sweden. In other parts of the world, Electrolux executives weren't so sure. This no-copying insurance policy would indeed make it difficult for an unscrupulous manufacturer to flood the market with a knockoff version of the robot. But the policy came with an expensive premium, and every Trilobite customer would have to pay it.

The marketing department got a vote in the final design of Trilobite. Around this time, fancy alphanumeric and graphic LCD screens were becoming popular in many high-tech products, especially cell phones. The marketing department decided Trilobite should cash in on the cachet these displays offered. Marketing specified an elegant display for the robot. "But Trilobite doesn't need an LCD screen," Per protested. "You just press a button to make it clean." Frustratingly, during the industrialization phase Per could only advise—and his advice was not taken. The fancy display was designed in, adding time to the schedule and cost to the robot.

The project suffered through yet another interruption. To address some inefficiencies in its global operations Electrolux decided to reorganize. The Trilobite project halted for a time while things were sorted out.

Altogether nearly ten years elapsed between kickoff and launch. But early in the new millennium, at last, all was in readiness. Half a decade after its TV premiere, Trilobite strode (rolled) boldly forward to claim its birthright as leader of the robot revolution. For the first time in history, customers could buy a robot vacuum.

CLOAK AND DAGGER

Electrolux caught our team flat-footed. In November 2001, after years of little to no news, the corporate giant from the land of fermented fish finally announced that Trilobite was now available for sale. The Roomba applecart was soundly jostled.

The introduction of a significant new competitor created much uncertainty. "How does this change things?" we asked each other. But initially no one could say. Plausible answers fell somewhere between the limits of "Not at all" and "Game over."

Contemplating the possible impact, we appreciated the extreme asymmetry of our situation. Electrolux was known globally, collected billions in annual revenue, and had a rock-solid reputation in the industry. iRobot—except for a few robot enthusiasts—was universally unknown, had never sold a vacuum of any sort, and held exactly zero credentials in the consumer floor-cleaning industry. If Electrolux intended to dominate the (thus far nonexistent) robot vacuum cleaner market, it appeared to be theirs for the taking. When we launched our product, would the Scandinavian super-Goliath even notice our baby David learning to crawl in a corner of the battlefield?

We scoured the internet and other sources, digging for scraps of information. A few tidbits began to emerge. Trilobite was round like Roomba, but taller, over five inches. We were planning to incorporate a side brush into Roomba to move otherwise unreachable dirt into the path of the main brush, Trilobite lacked this feature. It had sonar sensors, but their function and purpose were mysteries to us. Advertising copy suggested that Trilobite had an efficient way of covering the floor, but details were maddeningly scarce. We couldn't decide—had Team Trilobite found a clever/low-cost way to execute a boustrophedon path? Or was marketing verbiage just dressing up the same random bounce method Roomba used? And at last we learned Trilobite's price, about $1,500.

Given these clues, the waves of angst lapping our shores receded a skosh.

We had decided that a floor-cleaning vacuum needed to be short enough to fit under most furniture and standard kitchen toe kicks. Five inches was too tall. Maybe in Sweden the clearance under furniture and the height of toe kicks was more generous, but in the US, Trilobite's lofty stature would exclude it from many places we felt a robot vacuum ought to clean.

Sonar sensors typically measure distances in feet or at least inches. How, we wondered, did such a sensor earn its keep on a robot that had to clean right to the edge of all obstacles? What use was it to measure things that were far away?

We were also curious about the absence of a side brush. Wouldn't that leave an uncleaned strip between the end of the brush and the wall, an artifact due to the position and finite width of the drive wheels? Omitting the side brush would also seem to leave the dirt that collects in corners undisturbed. Or maybe we were missing something. Might the unrevealed underside of Trilobite harbor a clever cleaning mechanism we hadn't thought of?

Finally, the price left us scratching our heads. Did basic economics force Electrolux to charge such a high price, or was $1,500 just what corporate executives thought the market would bear? If Electrolux's costs were in the same ballpark as ours then we were in big trouble. They could easily reduce their price to fend off pesky competition from us. Our price was already as low as it could go.

These questions were critical for us. To answer them, we needed to dissect a Trilobite.

So began Mission Acquisition. It turned out to be more difficult than expected. First, Trilobite was not widely available. It seemed it could only be bought, in person, at three Electrolux stores in Stockholm. Unfortunately, our resources were few and we were unable to find a contact in Europe who could make a covert purchase for us in a timely fashion. So if the mountain wouldn't go to Mohammed … Winston and Jeff would fly to Stockholm and snag a robot.

We might have tried harder to get someone in Europe to send us a Trilobite, but we'd heard that Electrolux had set up robot display areas in their stores to demonstrate their machines. Seeing the robot in action, watching how store personnel used and explained it, and asking questions seemed as though it might prove very instructive.

On the long flight across the Atlantic, the intrepid travelers began to worry. What if Goliath is not complacent in his strength but instead is on high alert for potential Davids? It would be most imprudent to arrive at the Electrolux store and announce, "Hi! We're a competitor here to learn your product's secrets. May we purchase a robot, please?" Of course Winston and Jeff wouldn't say that, but what if the sales agent guessed?

Clearly some subterfuge was called for. Rather than simply walk in, buy a robot, and walk out with it as any innocent customer might do (imagine the alarm bells that would set off!) Winston and Jeff needed a plan. And in short order Jeff concocted one—a ploy so cunning and stealthy that is must surely evade the suspicions of even the most competitor-vigilant Electrolux sales agent. Because Jeff spoke fluent Spanish, he would impersonate a young, wealthy Argentine bon vivant, keen to be first to possess the latest technology. Winston would be his valet/translator.

Thus, cloaked in the verisimilitude of faux Latin American wealth and excess, Winston and Jeff visited every Electrolux store in Stockholm. At each venue they asked for a demonstration of Trilobite, they observed how employees operated it, and they asked many questions. The queries were the reason for visiting different stores. The plot might be foiled, they felt, if too many questions were asked in any one place.

At the final store, Jeff feigned increasing interest and excitement as the Electrolux employee described the robot. At last, in crescendo he declared, "I take two!"

The plan worked flawlessly. Winston and Jeff paid for their purchase (*not* using an iRobot credit card!) and departed the store with a pair of Trilobites in tow—the too-trusting Electrolux sales agents none the wiser.

Back in Somerville the Roomba team eagerly awaited the return of the travelers with their booty. Somewhat imprudently we were standing by with screwdrivers and wrenches in hand ready to dismantle a Trilobite the instant it appeared. Fortunately, a cooler head (probably not me) said, "Let's at least see how it works before we tear it apart."

So we read the instructions, set up the robot, and turned it on. The text told us that Trilobite would first follow the wall until it had circumnavigated the room. Afterward it would crisscross the interior until all was cleaned. It did that, more or less. We examined the robot's behavior carefully for signs that it was doing something different from a little dead reckoning and a lot of random bounce. But no, Electrolux had not beaten dead reckoning's fundamental problem, that positional errors never stop growing. Trilobite's inventors were working from the same robotic playbook we owned.

Trilobite had neither side brush nor any other mechanism for moving dirt into the path of the vacuum inlet. This meant that the robot would indeed leave a narrow, uncleaned swath next to walls and in corners.

Before dismantling the robot, we took a beat to admire the beauty and elegance of what Electrolux's engineers and industrial designers had wrought. The robot had a pleasingly smooth matte cinnabar finish. The trilobite-inspired grilles gave it a high-tech look. The carrying handle was seamlessly integrated into the design and the robot felt solid and sturdy in a way that inspired confidence. In a two-way beauty contest glamorous Trilobite would claim the crown while compassionate onlookers spoke consoling words about Roomba's great personality.[8]

When we finally got to take a look inside, we were all impressed with Trilobite's technology. The sonar was a wonder. The emitter wrapped 180 degrees around the robot giving it a cool-looking grille effect. Numerous tiny sonar receivers dotted the robot's front. We concluded that the purpose of the sonar was not to determine the range to distant objects but rather to give the robot a light touch. It was this mechanism that enabled

Trilobite to navigate a forest of goblets filled with wine and yet spill nary a drop on the carpet.

A sticker on the robot's bottom informed us that Trilobite consumed 90 watts of power (Roomba used 30). This high level of power consumption was likely responsible for Trilobite's weight, about 11 pounds—as described earlier, lots of power plus a reasonable running time equals a big heavy battery. The vacuum mechanism was cunningly designed to be as efficient as possible. But it was still a vacuum with a large inlet area, and that gobbled power.

Instead of our tiny microcontroller with its 256 bytes of main memory, Trilobite used the same sort of powerful microprocessor that was employed in early Macintosh desktop computers!

Designers had not bestowed the boon of cliff sensors on Trilobite.[9] In their place were mounted some number of magnetic-field sensors on the robot's underside. Accompanying Trilobite in its box, wound into a coil, was a long strip of magnetic material with adhesive on one side. The user was instructed to adhere a length of this material across the opening to any area where it would be unsafe or undesirable for the robot to operate. One function of Trilobite's magnetic sensors was to detect and turn away from such magnetic strips. (This was one of the confinement options we considered but rejected when developing Roomba's virtual walls.)

Trilobite included one helpful feature that we'd chosen to omit from Roomba's first iteration. Trilobite could recharge its batteries automatically. It accomplished this with the aid of a charging dock included with the robot. In profile, the dock was L-shaped. On the vertical wall were mounted two charging contacts. To access them, the robot drove up onto the thin lower part of the "L." Under this part, we gathered, were mounted magnetic strips that let the robot identify the charging dock and maneuver itself into the charging configuration.

We applauded using the same sensors for different purposes. That Trilobite's magnetic sensors served both to exclude the robot from certain areas and to identify the charging dock was parsimonious. But we were uncomfortable that the robot's safe operation depended on the user properly installing the magnetic strip. A sticker on the robot's top surface informed users of their obligation to mark stairs with the strip so the robot wouldn't fall. We had decided it should be the robot's job to keep itself safe.

Given the solid technology, the high quality of the product, and the fact that Trilobite was assembled in Sweden, it seemed likely to us that Electrolux probably did need to sell the robot for around $1,500 to make a profit. In 2001 the robot vacuum market remained unproven, so it was plausible that customers would willingly pay a big premium for a robot as elegant and cool-looking as Trilobite. But in our view, both robots had the same description ("Trilobite/Roomba is a little round robot that cleans your floors"), and worked about equally well (according to our tests). That being the case, we figured Roomba's $200 price would give us a significant advantage. So, for the rest of Roomba's development, we remained uneasily confident that Roomba would beat Trilobite in the marketplace.

TASSEL HASSLE

Some throw rugs have a decorative fringe at their edges. But the playful, almost festive appearance of this carpet bling belies a heart of darkest evil. The stringy menace of rug tassels turned out to be Sherlock Roomba's Professor Moriarity. If Roomba happened to encounter the tassels while driving *off* the rug, all was well, the main brush tended to straighten the tassels making them look neat and orderly. But if the robot approached from any other direction, the tassels would typically wrap themselves around the brush and try to pull the rug in after them. That usually left the brush hopelessly jammed and required the user to intervene.

The team tried to solve the tassel problem in many ways. We wondered if modifying the wire bale would help. But we could find no wire spacing or alternative bale design that prevented jams.

The core of the cleaning brush was a twisted wire from which the bristles protruded. The diameter of the twisted wire pair was quite small, so tassels tended to become tightly wound around the core. We tried adding a plastic structure inside the brush to make the effective diameter larger. That helped but didn't fully solve the problem.

Maybe another inventor working in a different domain had discovered the solution, we hypothesized. Possibly someone had designed a brush cleverer than any we'd considered. We cast a wide net looking for the holy grail of brushes, one that would stroke tassels without tangling them. Eventually, we found and purchased a promising product, a spinning motorized cylindrical hairbrush that advertised, "guaranteed never to tangle hair!"

A quick test of that bold claim seemed in order, so we asked our long-haired colleague, iRobot software engineer Jennifer Smith, if she would help. Too gracious to refuse (Jennifer was from Canada), she said she would. We handed her the brush. With some trepidation, Jennifer bravely pressed the button to start the brush spinning and then lightly touched it to the bottom of her locks. The brush instantly grabbed on and climbed to her ear while spinning her hair into an extravagantly tangled mess. All this in the milliseconds before she was able to release the button. Jennifer spent the next hour painstakingly extricating herself from that faithless brush.[10] After that there didn't seem much point in repeating the experiment with actual rug tassels.

If we couldn't find a mechanical solution, maybe a software trick would work. Built into the robot was an algorithm that continuously monitored the brush to check whether it was jammed (excessive current draw revealed the condition). We attempted to use that information to escape tassels— when the excessive-current alarm tripped the robot would cut power to the brush and then try to drive away. The idea was that moving the robot while leaving the brush free to spin would allow the brush to turn backward and unwind the tassels.

That worked sometimes. But depending on the length of the tassels, and which direction the robot drove away from the rug, the jam might persist. We wondered if spinning the brush slowly backward while driving away might be the best solution.[11] But we ran out of time before we could fully test that solution. We were forced to go into production without entirely foiling Moriarty. Roomba could negotiate short tassels but continued to struggle with long ones. Roomba's owner's manual advised users that if their rugs sported long tassels, they should fold them under the rug. Products always involve compromises.

ELEVEN OF TWELVE

One evening around seven or eight o'clock, after the rest of the team had left for the day, Phil and Eliot remained behind in the office, each working quietly in his own space. This was a frequent occurrence as work was always abundant.

Suddenly, with no warning, a deafening bang rang out. This was immediately followed by pieces of debris clattering to the floor around the office. Phil and Eliot leapt to their feet to investigate. Wisps of smoke hanging in the air quickly gave away the coordinates of the detonation's hypocenter.

Roomba was powered by a removable battery pack. The pack's dimensions were about 5.5 x 2 x 2 inches. Each pack was a thin-walled plastic box that contains 12 cylindrical, rechargeable nickel-metal hydride cells connected in series. Metal straps, spot-welded to terminals, connected each cell to its neighbor. As Phil and Eliot looked around the office, they saw the scattered components of a battery pack. Some cells were lying by themselves on the floor, some were still attached to their mates, and at least one, showing evidence of fire, had partially melted into the carpet. Phil and Eliot tallied as they collected. Eleven battery cells were accounted for.

Where was number 12? Eventually they managed to locate some small bits of torn metal and plastic mesh material. Apparently, cell number 12 had exploded, thoroughly obliterating itself in the process.

Further details were pieced together the following day. Chris Casey had been testing the battery-charging circuitry for Roomba and had left a battery pack sitting on his desk to charge overnight. To accomplish charging, his circuit employed a chip designed solely for that purpose. The chip was meant to give the battery pack ample current so as to charge quickly and then shut off when full charge was achieved. Apparently, the chip had neglected the shutting off part.

Small robots are nowhere near as dangerous as large ones. But even diminutive devices have their moments. Chris redesigned the charging circuit and Roomba's battery packs have remained inviolate ever since.

STOCKHOLM SEQUEL

Winston entered the Roomba office one morning to find—to his great vexation—that the second Trilobite robot he and Jeff had acquired in Stockholm was in pieces. Unbidden and unauthorized, an engineer not on the Roomba team had become curious about how Trilobite worked and decided to enlighten himself by dismantling the robot and leaving it in a heap. This was unfortunate because Winston had plans for that machine.

Easier and more reliable than trying to reassemble the robot would be making another trip to Sweden. Sara was offered a solo mission; she instantly accepted. The first part of her early winter trip went well. Sara flew into Stockholm's Arlanda airport and took a train to the center city. She entered the Electrolux store and, despite lacking the smoke screen of a faux personality, managed to purchase three Trilobites without arousing suspicion.

Afterward she enjoyed an art gallery, a visit to a then-exotic Ikea store, and the fascinating spectacle of birds fishing through holes in the ice of the city's frozen-over canals. Later, Sara began her return journey to Boston by boarding a flight that connected through Heathrow. It was during her layover there that trouble developed.

As she waited in the lounge area, an announcement echoed over the PA, "Will passenger Farragher please report to the departure gate?" Obediently complying, Sara was surprised and alarmed to be met there by two sternly unsmiling security officers. She blanched as they told her, "We've removed your luggage from the airplane."

It seemed that in the absence of Sara having done so, airport security had constructed a faux personality for her: Sara was a modern Mata Hari, intent on the destruction of the Western world.[12] Or so security's evidence suggested. This nearly inescapable conclusion was based on an X-ray of Sara's checked baggage. Could she, the officers demanded, offer an innocent explanation for the three large, menacing items stuffed with electronics in her luggage? No one in airport security had ever seen anything like them before.

While Sara was preparing for her trip, Paul Sandin had made a providential suggestion. As soon as she acquired the Trilobite robots, Paul said, she should take them back to her hotel room and make sure that they were fully charged. This she had done. Thus, when challenged at the airport, all three robots were prepared to demonstrate their peaceful capabilities.

Choosing her words carefully (could she be charged with industrial espionage she suddenly wondered?), Sara described the robots' innocent floor-cleaning purpose. The security officers remained impassive. Then she activated one of the robots, allowing it to begin cleaning the carpet near the departure gate. Although the agents retained their doubts about vacuuming robots, they allowed that these particular items did not appear to be the grave security threat they initially believed. After a 20-minute delay, Sara—with her robot-containing luggage restored to the cargo hold—was allowed to continue on her way.

What would have happened had the Trilobites' batteries been dead, preventing Sara from demonstrating that the robots were the benign servants as she claimed? Might the Tower of London have acquired a new occupant? Fortunately, that outcome remains speculative.

NETHERLANDS BEFORE THE DIKES

Every Tuesday iRobot held a staff meeting. Attendance was expected. Typically, each of the heads of the various divisions (there were 11 at one point) would rise to talk about the projects under their management—the progress being made and what would be coming up next. As a rule, only happy news was presented.

Winston usually spoke for Roomba. But he rarely said a lot and seldom got many questions. Our project, it seems, was not regarded as particularly sexy by some of our colleagues. Coworkers might be eager to share their suggestions concerning how we should design the robot but, beyond

Dave Nugent's early flirtation, there was no crush of volunteers jockeying to join our team. Winston liked it that way. He used to tell us, "Roomba is like the Netherlands before they built the dikes." The long-ago Netherlands were far from the center of things, there was a lot of mud, and scratching out a living took a great deal of effort. Residing in such a place appealed only to those few who understood how rich was the soil and how great was the potential. It was prudent to keep the region's prospects quiet because, had anyone had wanted to move in and take over, defending the marshy flats was near impossible.

The sentiments my old friend John had expressed to me recently in the lunchroom were popular ones. Some of the most desirable projects at the company tended to involve exotic technology and high-priced robots— like those for the military. A low-priced product that shunned technology and groveled in the dirt generated lesser enthusiasm. Everyone knew that floor-cleaning robots represented an old idea and that none who'd tried building one had ever made any money. The ignoble outcome of our Clean project likely reinforced that attitude.

We were happy in our relative obscurity. Eliot especially (after his dinosaur project) felt that fewer people looking over our shoulder was the more comfortable situation. We needed no adulation and were content to spend our days scratching in the mud, dreaming of the plentiful harvest to come.

But there was at least one downside to working on a project others considered unexciting. Like legendary comic Rodney Dangerfield, we got no respect. The way managers of some other projects saw it, Roomba didn't merit the resources we consumed—that money would be much better spent on the efforts they were leading. Regularly, at meetings of his peers, Winston was forced to repel attempts to poach Roomba's engineers and to defend the very existence of our project. Fortunately, Colin, Helen, and iRobot's board of directors were unmoved by such clamor and never entertained dismantling or diminishing Roomba. Roomba remained afloat despite the occasional broadsides of company politics.

BEEPS AND BOOPS

User and robot must exchange at least a minimal amount of information. The user needs to tell the robot what to do, and the robot needs to tell the user whether all is well. We called the system that accomplished this the Human/Robot Interface (HRI). And like everything else associated with Roomba, our HRI needed to be super-simple and dirt cheap.

The tell-the-robot-what-to-do part of the interface was, in principle, quite straightforward. Our ideal solution was a single push button labeled "Clean." Press that button and the robot would run until the floor was free of dirt. But there were complications.

When the robot cleaned a big room, it needed to run for a long time to make sure that every part of the floor was visited. In a small room Roomba could run for a correspondingly shorter time (reserving battery charge in case the user wanted to clean another room). And it should also adjust its running time depending how cluttered the room was. More clutter meant more bumping and escaping, and therefore a longer time to clean the same area. Unfortunately, Roomba had no sure way to figure out either the size of the room nor the level of clutter present.

So we reluctantly gave the time-estimating job to the user. Instead of one button, we supplied three. They were labeled "Small," "Medium," and "Large." Pressing any button initiated cleaning. But how long the robot ran before stopping was determined by which button the user pressed. Small yielded the shortest runtime, Large the longest, and Medium was in between. The three buttons were sure to cause some confusion; how is an owner supposed to know what size room qualifies as big, say? But we expected users would quickly figure out which button worked best in which room.

The output part of the HRI allowed for more possibilities. As noted, high-tech gadgets of the period were beginning to sprout LCD screens. These cool and versatile displays could deliver instructions, messages, logos—static or dynamic information of any kind. We admired Trilobite's beautiful backlit LCD display. It glowed a soft blue and showed both text and graphics. But of course, such opulence was not for us. We needed something affordable.

Attaching several labeled LEDs to Roomba's top might do the trick. Illuminating a label would inform the user of the robot's status, e.g., working smoothly or in need of assistance. But it didn't seem sufficient to use only LEDs. If something went wrong, users wouldn't know unless they happened to look at the robot—and, unlike a toddler, Roomba was supposed to function without being constantly watched.

Sound was an obvious candidate. Playing an alert sound would get users' attention and let them know right away if the robot needed help. So, we decided, Roomba should have a speaker. Those who demand high-fidelity end up spending big dollars to equip their stereo systems with fine woofers, tweeters, and mid-range horns. But if sound quality, efficiency, and power are only nice-to-haves, then a tiny speaker costing only a few cents will work just fine. We cared a lot more about cost than those other things.

Even with a crummy speaker it would have been possible to make Roomba play back complex sounds or tunes or even speak. Those abilities would require us to buy memory chips to store the sounds and another chip to play them back. We felt that the cost of such components exceeded their value—would playing Beethoven's Fifth make the floor any cleaner? So that was out.

The robot needed a microprocessor to control its motions—we had to buy one no matter what. And for no extra cost we could have the microprocessor generate sounds. All we needed to do was to take one of the processor's output lines, connect it to our cheap speaker through a cheap amplifier, and then toggle the line. If you turned the output line on and off 100 times per second, for example, the speaker would produce a sort of raspy 100 Hz[13] tone.

Suppose we wanted to make Roomba imitate a doorbell. That typical ding-dong sound can be made by playing the musical notes E and then C. The corresponding standard frequencies for those notes are 329.628 Hz and 251.626 Hz respectively. But now we had a problem.

It turns out that making music with our bargain basement microprocessor was a lot like trying to play an old piano where most of the keys have gone missing. Actually, it was even worse than that. Here and there on the piano you'll find a group of two or three keys that play harmoniously together. Several missing keys away you'll find another such group. But with our microprocessor the keys in the first group might sound dissonant when combined with those in the second. So, depending on what notes you want to include in your tune, it can become a maddening exercise to find enough compatible keys to make beautiful music.

The technical reason for this awkwardness is that our microprocessor couldn't generate just any frequency. One of our dirt-cheap chip's many limitations was that it could play only certain frequencies—and none of them corresponded exactly to standard musical notes. Our microprocessor could make any tone with a frequency of 9606 Hz divided by an integer two or larger.[14] Choosing a divisor of two and the speaker would sound a 4803 Hz tone. If we divided by say eight, we would hear a 1201 Hz sound. What if we wanted to play 1046.5 Hz, the C that's two octaves above middle-C? We were out of luck—the closest we could get was 1067 Hz using a divisor of nine.

But—unless you have perfect pitch—matching standard musical notes isn't all that important. It's much more critical to play sequences of notes where the notes have the right ratios to each other. If we look through all the frequencies the microprocessor can play, we can find notes separated by a musical third, or musical fifth, and so on. We can then use these ratios as the building blocks for short tunes.

That is exactly the challenge Paul and Phil pounced on. Phil started by building a spreadsheet showing all the combinations of notes the microprocessor could play. Paul and Phil then went through the chart trying out sequences of notes that would make pleasing alert sounds, like the tones Roomba played when it started up, or shut down.

The rest of the team endured a few weeks of, "Hey, listen to this," or "What about this one?", or "How does this sound?" followed by a few musical notes. For my money, the best thing Paul and Phil came up with was the pair of dissonant tones, the uh-oh sound Roomba makes when something goes wrong. (This sound plays, for example, when the robot halts because a wheel has dropped over an edge or when the brush is jammed.) After much diligent effort, the Rodgers and Hammerstein of the Twin City Plaza completed the full score of Roomba's HRI symphony.[15] Our mind-child was ready to beep, boop, and uh-oh its way into the hearts of its owners.

STEERSMEN

Examine the history of any great product—one that delights customers and exceeds expectations—and you'll inevitably find a harmonious confluence of marketing and engineering. Marketing that uncovered customers' sometimes unspoken wants, and engineering that satisfied those desires with parsimony and elegance. Ironic, then, that the relationship between marketing and engineering is traditionally fraught. Friction arises because each discipline has its own constraints and challenges that are not often understood or appreciated by the other.

Engineers demand that, from the start, they be given a complete and unchanging list of specifications. Without that, they feel like a hapless parent trying to pack the car trunk for a family trip—an exercise that becomes a nightmarish game of Tetris as the kids bring out items one at a time in no particular order.

Marketers for their part, demand that engineers identify what they can feasibly build and what it will cost. Absent that, marketers have no way to estimate customer interest or formulate a sales strategy. An awkward impasse develops: Engineer to marketer, "I can't tell you the feasibility and cost of the product unless I know the full details of what I'm building." Marketer to engineer, "I can't figure out the details until you tell me what you can reasonably build and how much it will cost."

Many mature companies often "solve" the problem by setting up an engineering/marketing collaboration. But in execution that alliance can become an endless series of meetings where, because of their different perspectives, parties talk past each other. Misunderstandings lead to frequently revised specs, changing specs lead to missed milestones and deadlines, and both lead to frustration.

Still, the process usually works. With intelligent and capable participants, with enough goodwill, discipline, and time, a successful product can emerge. The Roomba team possessed an abundance of all those critical ingredients, save one: time. We couldn't afford the temporal inefficiencies usually inherent in the product development process.

Enter Phil Mass.

Phil had ended up an engineer—a seeker of practical solutions to real-world problems. But he began in physics, a discipline focused on the discovery of fundamental principles. And his second major was sculpture, a thoroughly non-technical field driven by artistic creativity. Phil's interests encompassed pretty much anything he hadn't yet learned about. Polymath Phil even found *marketing* fascinating! Go figure.

The rest of we technical-team troglodytes might tolerate marketing as a necessary evil, to be endured for the sake of selling a few more robots, but Phil saw marketing as compelling in its own right. Marketing writ large was a window through which Phil could understand how a customer would use something he built. It could illuminate the issues that might give a customer problems and it could guide him toward implementing features that would delight customers. Like all of us on the Roomba project, Phil urgently wanted our robot to go out into the world and succeed. But much more than his technical peers, he was convinced that effective marketing could help.

Thus, a virtuous dynamic established itself. Phil and Winston would meet on a near-daily basis to develop marketing and engineering goals and to track progress toward each. Having helped establish all the team's goals, Phil understood and valued the synergy of the engineering and marketing aspects of the product. He spoke both languages and could translate in either direction. This let Phil act as a bridge between worlds. (Probably he was all the more effective in that role because he never told the technical team what he was doing.)

Sophisticated products are made up of myriad features. But on our robot especially, disparate features often didn't play nice with each other. Any given one had the potential to affect any other, usually in some complicated and adverse way. For example, if we chose to increase Roomba's cleaning aggressiveness, mobility might diminish—the robot would clean better but would get stuck more often. Or, if we say, added a convenient automatic recharging dock we might have to eliminate the virtual walls in order to pay for it. Exactly which features the robot should include, which would be excluded, and how they should be tuned relative to each other was our perpetual Gordian knot. Following the convolutions, appreciating the tradeoffs, and investigating possible alternatives consumed a large fraction of the team's effort.

Often, before events forced us to decide, Winston had carefully examined the issue in question and had worked through the tradeoffs with Phil. They compared the alternatives and picked the one that seemed to best harmonize customer desires and engineering feasibility. Then, during team discussions in the Roomba office, Phil would gently steer deliberations toward that choice. Rarely, some subtlety Winston and Phil hadn't fully appreciated led the group to a different selection, but generally the pair's choice prevailed. This sort of thing happened again and again. But, due either to Phil's exquisitely refined persuasive skills or my own immense obliviousness, I never caught on. Until years afterwards I thought the technical team had arrived at Roomba's design largely without marketing input.

Roomba's marketing and engineering imperatives appeared to align effortlessly. But, as when a skilled magician performs a mystifying trick, apparent effortlessness belies enormous unseen toil. Phil and Winston worked constantly. They often spent hours each day researching, discussing, and planning next steps. At the same time, Phil was responsible for writing and debugging all the robot's code and Winston had many duties beyond marketing and working with the technical team. The stress was considerable and both Winston and Phil paid a personal price.

But because they were young and driven and unencumbered by domestic obligations, an all-consuming obsession held romantic appeal. They welcomed their new robot overlord! And so, Phil suppressed his alarm when, during episodes where both ends of the candle blazed, he would lose handfuls of hair. Likewise Winston shrugged off the occasional morning when, looking into the bathroom mirror, he failed to recognize the person staring back.

With the weary but unseen hands of Winston and Phil on the wheel, Roomba steered a true course.

NOTES

1. Roomba came with a rechargeable battery. This focus group participant was telling us that she'd expect to pay $25 for a Roomba plus its battery. But that she'd value the battery without the robot at $50.

2. The team's failure to appreciate that customers would insist on a vacuum was the unforeseen customer expectation mentioned earlier in the "Helpful Heresy" section. Despite our mastery of floor cleaning, our technology-focused point of view failed to anticipate this aspect of the typical customer's mindset.

3. The vacuum's impeller was a specially shaped fan—basically a disk with perpendicularly attached, curved, rigid vanes. When the disk spun, centrifugal forces caused the air to be drawn in at the center of the disk and flung out at the edge. Roomba used a "dirty" impeller meaning that the air with entrained dirt flows through the impeller; a filter prevented the dirt from leaving with the exhaust.

4. Helen Greiner, as principal investigator on the PackBot project, proposed the concept and won the funding. Chi made it real. US Patent: US6263989B1.

5. DARPA—the Defense Advanced Research Projects Agency—has for years funded speculative ideas. It focuses on notions that have the potential to grow into useful products but are too risky for conventional funding.

6. The Navier–Stokes equations help a lot.

7. How *did* Winston's carpet make Roomba move sideways? We improved the robot's performance without fully understanding the phenomenon, but the carpet's nap played a role. Watch very closely as you step on any carpet. The nap will usually fold over rather than simply compress. Your foot or the robot will shift forward, backward, or sideways along with the nap. Winston's carpet was especially good at producing this motion.

8. *We* thought Roomba was exceedingly beautiful and that our industrial designer, Matt Cioffi, was doing a stellar job. But where Trilobite could afford metallic surface treatments and other fancy/expensive frills, Matt had only the shape and color of the plastic to work with.

9. Trilobite's inventors wanted to avoid robot tumbles as much as we did. So, they considered adding infrared-based cliff sensors. But they worried that these could be spoofed whenever lint and dust fouled the sensor. Dirt could not fool the magnetic sensors, so they were chosen. Roomba's cliff sensors *were* occasionally tricked by accumulating dirt, but our wheel-drop sensors supplied a fail-safe backup on the rare instances when this happened.

10. Jennifer ultimately forgave us for involving her in our imprudent experiment. She and I later shared an office and worked together on a very fun affordable vision project.

11. A later version of Roomba developed this method and greatly improved the robot's response to tassels.

12. The December 2001 date may help explain security's zealousness.

13. Hz is short for Hertz; it means "cycles per second.".

14 The geeky explanation for this limitation is that we generate the tone using what's known as a *timed interrupt*. At regular, selectable intervals, the microprocessor stops executing its main program and jumps to a very compact bit of code called an *interrupt routine*. This code either sets the speaker line high, sets it low, or does nothing. Then the microprocessor jumps back to whatever it was doing before. Turning the speaker on and off in this way creates a tone. But that tone can only be a quotient of the highest rate (when the line toggles every time the interrupt routine is called) divided by an integer divisor.

15 More recently iRobot added to new Roomba models the (now less expensive) electronics we were unwilling to pay for. Contemporary Roombas can and do speak. However, those original sounds that Phil and Paul developed, now iconic, are still used.

CHAPTER 14

Hour Eleven

T HE CLOSER WE GOT to launch the more the walls closed in. Home test-ing continued to uncover problems and complications but our options for fixing them narrowed. My go-to move of redesigning the entire robot every time we learned something new became untenable. We would have to write on an increasingly dirty slate.

CROSS-DISCIPLINARY TRIUMPH

The cliff sensors gave the robot a way to detect a precipice and avoid it. But should we rely on only that one system to keep the robot safe? We'd all worked with robots long enough to appreciate that the answer was no. No sensor is beyond reproach. All sensors seem adept at finding devious ways to deliver false output, at least occasionally. For the cliff sensors, we worried that a dust bunny might wedge itself beneath the sensor in such a way as to mimic floor presence when the actual floor had dropped away. And we suspected there were other failure modes we hadn't thought of.

A cliff sensor, reporting false data, could precipitate a damaging fall. To protect the robot and its owner from this possibility we needed to consider the details of how a fall would occur. The robot had three points of sup-port, one drive wheel on either side and a roller at the front center. If the robot drove along, not quite parallel with a cliff to the left of the robot, the left wheel would get nearer and nearer to the edge, and at some point fall over. The right wheel would continue driving forward. But because the left side no longer advanced, the right wheel would spin the robot around until the robot drove over the edge. A second scenario is analogous to this one, but with the robot obliquely approaching a cliff on the right. In the

 DOI: 10.1201/9781003540489-14

third scenario, the robot would advance toward a cliff head-on. Here, the roller would be the first point of support to drop over the edge. Afterwards, the drive wheels would keep moving the robot forward until it tumbled over as well.

In each scenario, the first step toward disaster was that one of the robot's support points dropped over the edge. If the robot could detect that event, it could command all motors to stop and prevent an actual tumble. Both drive wheels had springs that forced them down to prevent high centering. To implement tumble prevention, all we needed was to add a tiny mechanical switch that tripped when a drive wheel dropped down to the limit of its travel. (We called this switch the *drop switch* or *wheel-drop sensor*). But to detect scenario three, we needed to modify the roller.

Rather than fixing it directly to the robot, we mounted the roller on a carriage and confined that carriage to a vertical track. With this done, when you picked up the robot, the roller slid down the track an inch or so and then stopped. We positioned a microswitch that, like the drive wheels, tripped when the carriage/roller combo reached the limit of its travel.

Then we began testing.

The first problem was that occasionally the roller didn't shift down when it should. Our implementation was dead simple, there were no wheels or bearings, and the "track" was just a couple of pieces of plastic that slid past each other. With gravity as the only force, the pieces sometimes stuck rather than slid. To remedy this we added a second force, a compression spring that reliably pushed the roller down when the floor dropped away.

Paul and I did most of the testing and for that purpose Paul built a little three-foot by three-foot platform about seven inches tall intended to simulate a step. It was made of red oak, has a rounded edge just like a real step, and the wood was waxed and polished as it might be in an elegant home.

Our tests involved disabling the cliff sensors, positioning Roomba on top of the platform, and then making it drive toward the edge at various angles. All the oblique angles worked perfectly. When one drive wheel or the other dropped off the edge, the wheel snapped down to the limit of its travel, the wheel-drop switch tripped, and the robot came to a rapid halt—hanging over the edge but not falling.

But we found one situation where the robot regularly plummeted. When we forced Roomba to approach the step edge head-on it almost always failed to arrest its motion in time and so tumbled over the edge onto the floor. How to prevent this?

Tested in places other than our polished pseudo step, the robot generally worked. It seemed the step's (low but reasonable) friction has something to do with the problem. Maybe after the caster dropped over and the drive wheel motors turn off, the robot didn't stop immediately but slid forward a bit. That seemed likely because the bottom of the robot was made of plastic. That meant the friction between robot bottom and step was very small.

Rather than leave the robot/step friction to chance, we tried affixing textured, high-friction rubber pads to the robot's underside at strategic points. That helped. But not enough—Roomba continued to make unscheduled takeoffs followed by hard landings.

Paul and I called a meeting of the technical team. The five of us, Phil, Chris, Eliot, Paul, and I gathered around the meeting/worktable in the middle of the Roomba office. We needed to figure out how to attack this problem. Was it a mechanical, electrical, or software-based issue? Which discipline should be responsible for finding and implementing a solution?

Several critical systems had to work together to prevent Roomba from flinging itself from the heights. When the roller breached the edge of the step, the robot didn't immediately know that anything was wrong. First the roller had to drop down to the limit of its travel. At that point, the wheel-drop switch tripped. When the switch closed, the voltage in the wheel-drop circuit began to fall. Once that voltage fell below a preset threshold, the microprocessor was able to interpret that voltage as a digital "low," signaling that the switch had tripped. At some later point, the software would check that signal, decide that the wheel-drop switch had closed, and then command the drive wheels to stop turning. There was some inertia in the system, so the robot kept moving forward briefly even after the command. If the robot's center of gravity reached the edge of the step before all these activities had completed, Roomba would fall. The time between the physical event—the caster going over the edge—and the moment the robot stopped is called *latency*.

None of these things took very long. The latency in our system was probably only tens of milliseconds—literally less time than an eyeblink—but it was still too much.

As we talked it over, we could see that the problem was not entirely within the domain of any one discipline. Eliot said, "I can move the roller forward a few millimeters." That would give the robot more time to stop before its center of gravity reached the edge. Chris offered, "I can make the circuit react a little faster." That meant the microprocessor could get the message sooner. And Phil stated, "I can speed up the processing a little." That would stop the wheels a bit sooner. We retreated to our respective corners to make our changes.[1]

And it worked. When we added the contributions of all the disciplines, we subtracted enough latency to reliably bring Roomba to a pre-tumble stop. Roomba would not fall down the stairs even if its primary cliff sensor missed its cue.

For me, this cross-disciplinary triumph was Team Roomba's finest hour. It was one of the most satisfying episodes during the entire development. No manager held anyone's feet to the fire. We were all strongly motivated to find a solution. No one took the position that the work they'd already done was correct, so the problem must be someone else's fault. We all contributed to solving the problem. In the end, Roomba's design changed: the rubber pads were added and the mechanical, electrical, and software systems were all enhanced. Just a few days passed between the discovery of a significant and at first perplexing problem and the implementation of a reliable solution. This sort of cooperation exemplified our team's process.

PHIL'S TINY RAM

As mentioned, we deliberately selected a low-end microprocessor[2] to control Roomba. We had two reasons for this: (1) cost and (2) complexity. First, a chip with minimal computational power costs less than one with more. Second, we believed that entrenching a chip with a modest feature set at the heart of our design would discourage mission creep. As we started Roomba, Paul and I were still smarting from our Clean experience where the robot just kept growing bigger and more complicated. We enlisted the microprocessor to help Roomba avoid a similar fate.

Perhaps the most limiting aspect of the chip we chose was that it had only 36 input/output (I/O) lines. Each button, switch, or sensor on the robot used up one input line. And each motor, indicator light, or speaker used up from one to several output lines. Just connecting the basic functions of Roomba to the microprocessor exhausted all the available I/O lines. That meant that if later someone wanted to add a bell or whistle—no matter how cool it might be—there would simply be no way to connect it. Attempting to do so one would find that on every window into the microprocessor was taped the sign, "Closed, next window please."

My declared intention of restricting Roomba to a modest microprocessor was one of the features that attracted parsimony-minded Phil to the team—he was excited by the challenge of implementing a version of Rod Brooks' behavioral robotics in a tiny processor. But given the tight group we ultimately assembled, locking in a low-end microprocessor proved an unnecessary defensive measure. We were of one mind, viewing the similarity of our task to packing for a trip to the moon. Even one ounce of unnecessary weight might mean we'd never arrive.

The negative consequence of choosing a feeble microprocessor is that it made more work for the programmer. More capable chips included hardware that made certain events happen effortlessly just by writing a number or two to a special memory location.[3] Not our chip. It offered no shortcuts. Every operation had to be spelled out in full in software.

Another severe limitation was memory. The chip Roomba used boasted 256 bytes of RAM[4] and 16,383 bytes of program or ROM.[5] If a modern computer has an amount of RAM memory represented by the size of an Olympic swimming pool, then Roomba's memory was equivalent to a shot glass,[6] literally. Making do with such a tiny aliquot of memory requires heroic efforts.

To aid in his Herculean programming task Phil made a spreadsheet. In it he essentially kept track of every byte and sometimes every precious bit of his RAM. He also tracked how every byte of ROM was used. Challenges grew as time went on. Near the end of our development, space became so tight that every change to the program required a tradeoff. We weren't able to add a new feature unless we first eliminated an old one or figured out how to achieve the same functionality while using less code.

The chip we selected could be ordered with different amounts of onboard ROM at different prices. Phil had started with a 12,287-byte version. He did quite well, cramming all the needed functions into that constrained memory space. But late in the game we realized that we needed to add a *factory test* to the code. Accessed in a special way, the factory test code would perform a self-diagnosis of each freshly assembled Roomba after it came off the production line.

This broke the camel's back, forcing us to switch to the chip with the next larger memory size (16,383 bytes). Suddenly, memory space felt abundant. After Phil wrote and installed code to perform all the specialized factory tests Roomba needed, he found that unused code space remained. Letting nothing go to waste, Phil wrote a couple of *Easter eggs*. An Easter egg, as you may know, is an unpublished bit of functionality included in a product just for fun. If you hit the S, M, and L buttons and the bumper in just the right arcane sequence, you could turn Roomba into a musical instrument using the musical system Phil and Paul created to enable the sound effects. When the Easter egg was activated, each sensor would play a different note, so if you tapped the bumper and blocked the cliff sensors in the right sequence you could play a tune. Another Easter egg made Roomba dance a little jig.

ZEUS DEFENSE

Science museums often demonstrate the properties of static electricity with an engaging exhibit featuring a Van de Graff generator. These imposing machines, sometimes over 20 feet tall, incorporate a central nonconducting, hollow column topped by a smooth metal dome. When powered on, a belt running inside the column ferries electric charge from bottom to top. Within seconds, an enormous voltage can build up on the dome. A grounded object coming within even a few feet of the charged dome can initiate a terrifying lightning bolt that crackles and forks as it leaps between dome and object.

Something similar can happen with Roomba.

The lightning bolts Roomba generated were much smaller, but the principle was the same. Spinning the cleaning brush against a carpet created a static charge in just the same way as would shuffling your feet across the rug on a dry winter day. If you then touched a metal doorknob you'd see—and feel—the charge rapidly dissipating. The museum's Van de Graff generator might build up an electrical potential of over a million volts to produce a yard-long bolt. But if the spark jumping between your finger and the doorknob was, say, a quarter-inch long, it meant you were likely charged to over 6,000 volts![7] Why wouldn't that be fatal? Because the electrical currents and total energy involved are much too small to cause you any real damage.

But the situation is different for sensitive electrical components. The transistors, diodes, and other elements embedded in the integrated circuits that made Roomba work were literally microscopic. If one of these tiny parts was expecting to see, say, five volts and instead was momentarily exposed to 5,000 volts, or even 500 volts, it would be likely to expire on the spot, never to compute again.[8] This phenomenon, called *electrostatic discharge* or ESD, posed a significant problem for Roomba as it does for most types of portable electronic equipment. If the charge generated by the brush happened to find its way to an exposed bit of metal, it could travel to spot far from the brush. And that could lead to a tiny, but no less destructive, lightning bolts sparking off anywhere within the robot.

Chris Casey bought two pieces of equipment to help us diagnose and mitigate Roomba's ESD issues, an *electrometer* and a zapper (aka an ESD gun). Both were great fun to play ... or rather to conduct important experiments with. The electrometer enabled us to measure static charges that built up on Roomba or on the carpet that Roomba cleaned. We marveled that after Roomba had passed over, an electrical potential in excess of 20,000 volts was sometimes left in its wake!

Briefly I toyed with the idea of using this residual electrical field to build a localization system for Roomba. If we mounted an electrometer on the robot, I thought, maybe that could tell us when the robot had arrived at a spot it had previously cleaned. Unfortunately, that proved problematic as the charge on the carpet dissipated quickly (especially in humid conditions) and non-carpeted floors barely charged up at all.

Our problem with static electricity wasn't academic. It bit us several times during development. One iteration of the robot was able to clean a carpet all day long without incident. But if Roomba then trundled across a metal threshold a spark somewhere inside the robot caused the processor to crash immediately. That sort of thing was unnerving. It meant that testing the robot successfully for long periods wouldn't necessarily uncover all its flaws. You had to test it in just the right way. But the right way was seldom obvious.

Another time, when we thought we'd solved all our ESD problems, we tried saving money by making a small change to the robot. We switched the anti-throw rug ingestion grille across the cleaning mechanism (the bale) from metal wires to a less expensive plastic structure. Instantly, our ESD troubles reemerged. We'd thought that an insulator (the plastic) should work better than a conductor (the wire) in this situation. But since that was not the case, we went back to metal.

We addressed any ESD problems we discovered or suspected by repositioning wires and other metal components and by strategically wiring in protective circuitry. This circuitry included special components that automatically shorted high voltages to ground before they reached more susceptible elements. Once we had what we thought was a good solution we attacked the robot with the zapper. This let us apply little lightning bolts of any strength anywhere on the robot. When the robot successfully resisted all such attacks our confidence increased that it would survive the real world.

FOLLOWING WALLS

As described in Chapter 7 "Coverage," we'd learned a trick that improved cleaning efficiency in cluttered spaces—sometimes the robot should follow the wall. What counted as a wall? Pretty much anything—the actual wall of a room, a shoe, the platform of a rolling chair, the skirt that hangs down to the floor from some couches. The last example shows us that there are two types of walls, solid walls that will stop the robot no matter how hard it pushes, and flexible walls that the robot can drive through if it pushes hard enough.

The already-described bumper gave us one way to detect and follow a wall—arc forward to the right, upon collision, spin in place to the left, then arc forward again. We called this method *bump-following*. Bump-following can follow any sort of obstacle but hugging the wall tightly necessarily made the robot's progress very slow (because of all the spinning back and forth).

Bump-following also had long-term reliability implications. Spinning a motor alternately forward and backward created stress that decreased its longevity. Much better was to spin only in one direction as much as possible. Bump-following forced the outside wheel—the one farther from the wall—to constantly reverse direction.

Since following walls was something the robot would do very frequently, we wanted it to be efficient. It turned out we already had the technology in hand.

Suppose we took one of our cliff sensors, turned it on its side, and mounted it near the edge of the robot. When the cliff sensor was brought near a wall it would now detect the wall—before the robot collided with the wall. Since the robot knew about the wall in advance it could speed up the inner wheel and slow down the outer wheel, thus arcing away from the wall without having to reverse the direction of either motor.

This became our solution. All we had to do was to modify the cliff sensor's geometry a bit to optimize it for wall-following. Now instead of a pattern of arc, spin, arc as the robot repeatedly collided with the wall and turned away, Roomba executed a subtle waggle-waggle as it moved parallel to the wall without ever touching it.

PHIL'S TINY FLOOR

It's an understatement to say that autonomy is tough. The Roomba performance feature that gave us the most intense headaches for the longest time was the robot's ability to move around and not get stuck. Everything affected autonomy. The nature of the floor surface was a big one. And at first, any deep-pile carpet could freeze Roomba in place. So could thresholds between rooms and the raised tracks that guide sliding doors.

The floor-mounted, louvered vents that deliver warm or cool air to a room grabbed the wheels of early Roomba prototypes almost every time the robot rolled across them. Because of that we had to make Roomba's wheels almost four times wider than we'd originally planned.

As Roomba sampled a broader range of venues during testing, we repeatedly encountered floor and room features that called for changes to the mechanical system. We increased the force on the springs pushing the drive wheels down, altered the undercarriage, and reshaped the bumper among many other things.

But once the mechanical design became robust—giving the robot the *potential* to drive away from just about any pernicious situation—it fell to software to realize that potential. That was Phil's calling, the thing that had inspired him to come to iRobot in the first place. And by the culmination of his work, Phil had created a responsive behavioral control system in the best tradition of Brooks' behavior-based paradigm. He built a layered system where a collection of simple software modules (behaviors) combined to solve the complex floor-cleaning task.

Roomba's lowest-level, lowest-priority behavior was called Cruise. Cruise ignored all sensor inputs and offered only one suggestion to the robot: drive forward at one foot per second, forever. This the robot did, as long as no other behavior spoke up.

But, waiting in the wings, other behaviors were listening for their cues. One called Bump, with a priority one step higher than Cruise, watched the sensors connected to the bumper. When Roomba's bumper touched something, the Bump behavior triggered. Bump then commanded the robot to spin in place a random number of degrees. Since it had a priority higher than Cruise, as long as Bump was active the robot did what Bump ordered. Then, as soon as the robot had turned the number of degrees Bump specified, Bump became inactive. This let Cruise once again win the battle for control, and the robot resumed driving straight.

From the simple interaction between Cruise and Bump a more complex global behavior emerged. Roomba would move about the room, bouncing away from obstacles until the floor was clean. Or at least it would have had the world been a place friendly to floor-cleaning robots. In actuality, there were many more challenges Phil's code had to respond to—things like cliffs, virtual walls, and rug tassels.

Phil managed to satisfy all the requirement by programming a hierarchy of about 10 behaviors. Together they were generally effective at guiding Roomba through its cleaning task. But further testing revealed a new issue.

Often, when the robot wandered into a small space with a narrow exit, Roomba became trapped—bouncing repeatedly but never finding the way out. Unfortunately, such spaces formed frequently, especially around dining-room tables where a forest of chair and table legs conspired to create a virtual cage for Roomba.

Here Phil's simple Bump and Cruise behaviors weren't enough. They left Roomba ricocheting pointlessly unless it happened to turn in just the right direction at the right time to thread the exit. We expected customers wouldn't like so much unproductive activity any better than we did.

Phil made it his mission to find a solution. First, he asked Paul to build a miniature test pen and populate it with pegs to simulate chair legs. To make the setup easily reconfigurable, Paul mounted the round pegs on thin square supports. He attached Velcro to the bottom of the supports so that they would stick to the carpeted floor. When Phil received the setup, he began working through different arrangements of pegs until he found a configuration that confounded Roomba. And then he repeatedly revised his code until he zeroed in on an algorithm that let Roomba escape reliably.

That took weeks. I remember well Phil sitting in his cubicle day after day, hands on his keyboard, leaning over the edge of his desk peering down at Roomba as it bounced around its little test pen. Again and again Phil would modify his code, load it into the robot, and then test its response.

The loading part was a pain that demanded saintly patience. The only way to reprogram our budget-priced microprocessor was to "burn" code into an external ROM chip. The process began by erasing the previous program. This required exposing the chip to ultraviolet light for about an hour in an instrument designed for the purpose. Then the erased ROM chip was installed in our standalone "ROM burner." Next Phil attached his computer to the ROM burner, uploaded his program, executed the burning process, physically removed the chip from the device, and then installed it in the robot. But wait, there's more! Installing the chip meant that Phil had to disassemble the robot, remove the processor board, and physically pry out the chip with the previous version of the code, and then press the newly programmed chip into the vacated socket. Then he could replace the processor board, reassemble the robot, and finally conduct his test. No matter how small the change, all this had to happen each time Phil updated his program.[9]

Later versions of our microprocessor offered easy, in-situ programming. But the cheap version we used required us to prove our love each time.

Phil ultimately solved the trapped-Roomba problem. He did this by adding a new top-priority behavior we'll call Houdini. Like Bump, Houdini watched the bumper sensors but was much slower to act. Only when Houdini noticed that the bumper had repeatedly hit something within a short period of time—as would happen when Roomba was trapped in a confined space—would Houdini take over. Then it would cause the robot to move forward as it hugged whatever object the bumper was touching. Keeping the bumper mostly pressed in this way let the robot shimmy through openings only a hair wider than the robot itself. Once Houdini guessed that it might be clear of the opening, it would rotate the robot by a specific angle—one painstakingly found through trial and error—that gave the robot the best chance of finally departing the trap.

STASIS

The team girded Roomba with sensors to detect many anomalous conditions. Among other things Roomba knew if its drive wheels were supposed to be turning but weren't. It knew if the main brush was spinning or jammed. And it knew if the bumper was pressed against an obstacle and if it was about to drive over a cliff.

But could our sensors detect *every* problem Roomba might encounter? We knew they could not. There were circumstances that led Roomba to believe that all was right with the world when in fact it was stuck.

Think back to the long-ago demonstration at the SCJ employee's house in Chicago. There, a rogue floor nail reached up and grabbed Roomba from underneath. Roomba couldn't tell that anything was wrong. The floor was slippery enough that, although the robot was motionless, the wheels just kept spinning, slipping against the floor. You may have experienced a similar situation while trying to drive a car up an icy hill—your speedometer tells you your car is moving when, in fact, you're stationary, just spinning your wheels.

In the Chicago scenario, Roomba's encoders told it that its wheels were turning, the bumper detected no collision, and the cliff sensors saw a solid floor; nothing seemed amiss. But the robot was in *stasis*—making no floor-cleaning progress. That was why I had to interrupt the Chicago demo to rescue Roomba.

Did such a situation doom Roomba to spin its wheels until the battery gave out, or might there be a way to recover? Recovery turned out to be possible, and once again it was random to the rescue.

To jolt Roomba out of any ill-advised complacency we created a virtual *anti-stasis sensor*. When Roomba moved about a room, it expected to encounter obstacles from time to time, and occasionally come upon a cliff. That is, the bumper and cliff sensor would occasionally activate. In software we created a timer that normally counted up, but we forced it to reset to zero whenever one of these sensor events happened. Anti-stasis sensor code monitored that timer. If the timer counted up to preset value, maybe 30 seconds, without being reset, Roomba would execute a random motion—spin in place a random number of degrees.

In normal operation the stasis sensor rarely activated. Roomba almost always bumped into something or saw a cliff before the 30 seconds were up. But if it was caught up on something like the rogue nail, there was a good chance that a random turn would set it free. Unfortunately, we hadn't yet thought of the anti-stasis feature at the time of the Chicago demo.

If Roomba were assigned to clean an especially large room—where it might travel in a straight line for more than 30 seconds—the stasis behavior could trigger even when there was nothing wrong. But being overly cautious caused far fewer problems than not being cautious enough.[10]

VROOM, VROOM

As will be clear by now, over the two-plus-year course of the project we built *many* different prototypes. Sometimes just one or a few, sometimes several at once. The demand for robots was constant. We needed them in-house to test different aspects of the design, we needed to hand them out to employees for home testing, and marketing needed them to show to potential vendors.

Since the Big Light incident (when Colin wanted to put an LED on Roomba's underside to point at dirty spots) he hadn't offered any further feature suggestions. But at some point, after watching a new prototype, he did insist on one. "Make the vacuum turn on first," he said.

Roomba ended up with five motors: two for the drive wheels, and one each for the main brush, the side brush, and the vacuum. Up to this point, as soon as the user pressed a button to start, the robot's program activated all five motors at the same time. The order of activating the motors, whether simultaneous or staggered, had no effect on the robot's ability to clean. But there was another issue that the engineering team was less adept at noticing: customer perception.

The way we had initially programmed the robot, the vacuum's sound was not identifiable amid the other noises the robot produced—the main brush consumed the most power and made the loudest racket. We'd gone to extraordinary lengths to add the vacuum to the robot at the last minute just so we could list it on the box. So, when we thought about it, it did seem prudent to reinforce the fact that the robot possessed a vacuum by letting the user hear it. We implemented Colin's idea (it took only a line or two of code) and the robot indeed sounded better.

REMOVABLE BRUSHES

We had regarded the ability of users to remove Roomba's brushes for cleaning as a "nice to have" rather than a necessity. Then Roomba went home with Nancy Dassault.

Nancy was iRobot's PR wizard and, as such, she needed to be familiar with Roomba's operation. We gave her a Roomba prototype to try at home. She brought it back the next day—it had stopped working. Nancy owned a pair of dogs who enjoyed indoor living as much as any human. During its brief run, Roomba discovered abundant evidence of this happy cohabitation. When we turned over Nancy's robot to examine its brushes we found a furry mess. Rather than nylon bristles, it looked as though our brushes were constructed of dog hair—the bristles totally obscured beneath the tangle of collected fur.

We were all appalled, Eliot most of all. We just hadn't envisioned that hair would create a problem of such magnitude. Suddenly our idea that it was OK to permanently install the brushes was revealed as painfully naïve. It might take a user hours to disentangle all that hair by teasing it from the bottom of the robot using the little brush-cleaning tool we planned to supply. To clean the brushes easily and properly they needed to come out.

Ideally, we would achieve easy-out, easy-in using a brush assembly held in place with a convenient clip or spring tab. The user would just press the tab to release the brushes. But that idea was triaged away. We discovered the problem a mere 72 hours before the deadline when we'd agreed to deliver our how-to-build-Roomba CAD files to the factory in China. There was simply no time to design and integrate into the robot a user-friendly brush-release mechanism. Instead Eliot plunged into designing the best compromise that presented itself. His solution was to add a new, removable plastic piece that retained the brushes. The embarrassing aspect of this last-minute design was that to remove the brush-holding part required a screwdriver. That made maintenance more awkward for our users, but it was the best we could do with the time we had.

ONES THAT GOT AWAY

As mentioned, triage was a constant necessity. We had time and cost constraints—the design had to be completely finished by a particular date, and the total parts cost of the robot had to be below $50 (I think we actually hit $52). Any feature that pushed us over those limits had to either be done in a simpler, cheaper way or eliminated. A lot of things were eliminated.

A notable example is Roomba as pyromaniac—where the robot drives into and then out of a fireplace spreading conflagration. We considered several options. One possibility was to equip Roomba with heat or flame sensors.[11] When the sensor activated, Roomba would turn away. But beyond the added cost was another even bigger problem, *false positives.* These happens when a behavior is activated by something other than the situation the sensor is supposed to detect. When Roomba pointed toward a hot oven in the kitchen or a wood stove or a fire burning safely in a raised-hearth fireplace, the robot would be likely to turn away.

False positives would degrade Roomba's performance in many common situations for the benefit of preventing it from behaving badly in one very rare situation. Ultimately, we decided to fix the problem not with sensors and software but with ink. We placed this sternly worded warning in Roomba's Owner's Manual: "Do not use this device to pick up anything that is burning or smoking, such as cigarettes, matches, or hot ashes."

The mischievous carpenters who built my house back in 1956 apparently decided to a play a very forward-looking prank on me. They set the height of the toe kicks beneath my kitchen cabinets at about four inches on either end then had them gradually decrease to around three inches in the middle where the counter turns a corner. This created a perfect trap for Roomba. While cleaning my kitchen floor the robot would begin following the cabinet where the toe kick was high and proceed to where it was lower. The overhang would thus very gradually descend on Roomba's top. The robot believed that all was well until it became wedged under the counter and stopped moving. Roomba was rarely able to extricate itself when this happened. We encountered the same problem under certain tables that included decorative near-the-floor, shallow arches.

Roomba was vulnerable to any hazard it couldn't detect. So, to fix the descending overhang problem we would have to add new sensors to Roomba. The bumper would need to detect not just collisions from the front and sides but also when something pressed down on it from the top. That was tough. Eliot had already taken heroic measures to get the bumper to work in two dimensions, adding a third would be even more complicated and costly. We decided to not worry about overhangs. I ended up sometimes putting a two-by-four under my counter so Roomba wouldn't get stuck.

Hair bedeviled Roomba. It could become wrapped around the brush requiring cleaning. But hair's most insidious trick was when it wrapped around the small, 0.1-inch diameter axle about which the brushes turned. Phil suggested, "Maybe we can design a way to expel the hair from the end of the axle." I said, "So you want to build a hair pump?"

But we never managed to build that pump. We had to content ourselves with tweaks to the design that reduced but didn't eliminate the problem. In hairy environments users needed to clean the brushes and axles more often than we wanted.

NOTES

1 We unanimously rejected one simple solution that we know would work. We could have reduced Roomba's cruising speed. Going slower would indeed have kept Roomba from tumbling but at the cost of impairing its overall performance.

2 We used a 44-pin, 40 MHz processor from Winbond, model W78E54B.

3 A popular example of this feature is built-in *pulse width modulation*. Simply issue a single command to make an output line toggle at a frequency of your choice, with a high or low duration also of your choice.

4 RAM stands for random-access memory. That's where Roomba put all dynamic information—all the variables in all its behaviors, all the values the sensors reported, the time until an event occurred, and so on.

5 ROM is read-only memory. All the code for running Roomba was stored in ROM.

6 I recently bought a computer with 16 GB of RAM memory. That's about 60 million times as much as Roomba had. An Olympic swimming pool holds 2.5 million liters of water, a shot glass about 44 milliliters. That's also a ratio of around 60 million.

7 The relationship between voltage and spark length is mediated by the *breakdown voltage* of air. Along with pressure and humidity, breakdown voltage also depends strongly on the shape of the electrodes, i.e. your finger and the doorknob. In the situation described, a rough approximation tells us that 25,000 volts will generate an inch-long spark. Ouch!

8 In practice, it was more likely that an internal spark would only cause the processor to crash. Crashes happened frequently during development. We killed microprocessors only occasionally.

9　Phil didn't have to wait an hour for each reprograming. The ultraviolet eraser accepted several chips at once. This let us keep a supply of already-erased chips on hand. Also, Phil usually left the top cover off his test robot to minimize disassembly/reassembly time.

10　In later versions of Roomba, engineers implemented a physical stasis sensor. The front roller was made half black and half white and a little light sensor was directed to watch it. As long as the robot moved, the passive roller spun, and the light sensor saw regular changes in brightness. But if anything halted the robot's motion the brightness changes would stop, and the robot would recognize that it was in stasis. This sensor enabled a significant improvement in robot performance.

11　Differential *pyroelectric* sensors are a type of heat-sensitive motion detector often used to automatically turn on outside lights. Ultraviolet sensors can be used to detect flames. Those were the types of senors we considered for Roomba.

CHAPTER 15

Um, Actually …

WE NOW RETURN TO the existential crisis recounted in the Prologue. It was spring 2002, and with everyone working frantically things were falling into place for Roomba's September debut. But what sounded like a lot of time wasn't. Launching a product required a huge effort and the coordination of many moving pieces. From now through launch, every day was scheduled, and every movement choreographed.

Most of Roomba's parts were made of injection-molded plastic. As previously noted, the first step in fabricating such a part is to use a CAD program to specify the exact shape and dimensions of the part. For the past few months, Eliot had abandoned any semblance of work–life balance in his effort to complete or revise all of Roomba's many parts. Once we were satisfied that every part was designed correctly and worked satisfactorily, we could deliver the CAD files to the factory. Then, in a step called *commit-to-tooling*, we would authorize the factory to commence with their part.

The factory's role was, first, to fabricate the molds that would make Roomba's parts. In this step, factory machinists would use Eliot's CAD files to specify the intricate shapes to be cut into massive chunks of steel. Those cavities in the steel would become Roomba's parts when injected with molten plastic during the molding operation. Machining the steel was exacting, time consuming, and expensive. We had planned for the factory to have just enough time to finish making the molds, molding the parts, assembling and testing the robots, and shipping them by sea from the Port of Shenzhen to the US to rendezvous with our launch date.

DOI: 10.1201/9781003540489-15

If we delayed authorization, everything might begin to unravel. But it would be even worse if we committed and then later decided to change the design. Anything more than a minor tweak to a part or two would require throwing away expensive, just-finished molds and starting over. Cutting the new mold or molds would almost certainly delay our launch. That in turn would upset the publicity, logistics, and vendor agreements we'd made. In the very worst case, a delay could exhaust our cash, killing Roomba and maybe the company.

Things had been marching along smoothly. The team had given our assurances that Roomba was ready for prime time—because we were certain it was. Then Winston asked us to do a quick little test. And Roomba failed.

There was no sugarcoating the disaster we'd just witnessed. We presented Roomba with some unchallenging, easy-to-pick-up dirt. But Roomba passed over it, leaving the mess behind as though it were some other robot's problem. Earlier we'd tested Roomba prototypes extensively, comparing and cataloging how much dirt Roomba could pick up relative to manual vacuum cleaners. Roomba performed well then; why had it suddenly abandoned its devotion to hard work?

The gravity and urgency of the moment keenly focused our minds. Meticulously we checked the robot. Was this particular prototype broken in some way? It was not. Was the main brush spinning as it was supposed to? It was. Did the pivoting carriage that held the brushes press against the floor with the proper force? It did. Had we changed the cleaning mechanism significantly from the last iteration? No—at least not that we were aware. We concluded: the robot was constructed as we intended, its components were functioning as they should, we were operating it in the correct way … it just didn't work. After our checks, we were no closer to a resolution than when our exercise began.

As we all stood there staring at the robot, unsure what to do next, ever-helpful Chris turned to me and said, "We'll be needing a solution for this problem. Shall we say, Wednesday afternoon?"

Chris was being facetious but not inappropriate. As the team's resident physicist, it was generally my role to supply a quantitative model for any physical phenomena we needed to understand in order to make the robot work. An insightful model explaining Roomba's unwelcome behavior would make all the difference right now. But I was flummoxed, I had no idea why Roomba wouldn't clean. Our tests appeared to have eliminated every possibility other than a serious but subtle design flaw. That flaw could be anything from an easily correctable oversight to a mistake the size of *Titanic*'s iceberg. We couldn't begin making the molds, because we had no idea how extensive a change would be needed to fix the problem.

It's embarrassing to admit at this stage, but we never had a quantitative understanding of the cleaning mechanism on which Roomba relied. Instead we had designed by analogy—making sure that the cleaning machinery we built into each robot strongly resembled that found in contemporary carpet sweepers. Although the omission had always troubled me, our lack of a sound, well-tested theory of cleaning never posed a problem before. The mechanisms we built all worked acceptably well, and there was always plenty of other work needing attention. So, the right moment to perform an in-depth study of carpet-sweeping mechanisms never arrived. The cleaning wasn't broken, so I didn't fix it. But now it was. And I felt largely responsible, worried that my lack of rigor had put our precious mind-child in grave jeopardy.

A little more was going on than the team appreciated. It was not a whim that brought Winston to the Roomba office that day. Rather, Winston's pop test was inspired by a problem marketing had discovered. As part of our launch preparations, our faux bon vivant Jeff Ostaszewski had been working on a promotional video. The idea was to position Roomba on a dirty floor in a typical looking home and then show how spectacularly well Roomba cleaned. But it wouldn't work. Using the same generation of prototype we'd just tested Jeff had been trying for days to show off Roomba in a good light. But his robot wouldn't reliably pick up dirt either.

Jeff and the rest of marketing had alerted Winston to the increasingly urgent problem. And Winston (and Phil) had tried to convey to the team that something was amiss. But the message hadn't gotten through—I have no clear memory of Winston informing us that Roomba wouldn't clean. A straightforward escalation would have been for Winston to simply walk into the team office and proclaim, "Marketing says Roomba doesn't work, you have to fix it."

That would have been problematic because we knew Roomba *did* work. For over two years we'd poured everything we had into making Roomba work. We'd resolved dozens of knotty problems and completed hundreds of tests that proved Roomba worked. We were very proud of what we'd achieved. In our minds we had nailed the cleaning problem. Why would we revisit it now?

Had Winston taken the obvious approach and just told us to fix the robot, our initial reaction would have been to pooh-pooh marketing's obviously misguided claim. No doubt, we'd have flung the problem back at them while reciting every engineer's favorite four-word phrase, "You're doing it wrong!" What we would not have done—and what Winston knew we needed to do—was to begin working immediately to find a cure for Roomba's malady.

So, almost casually, Winston asked the team to do a simple, impromptu test. The scenario then played out just as he had intended. By our own hand, we were confronted with undeniable evidence that the robot wasn't working, and we felt a profound compulsion to figure out why.

It was an all-hands-on-deck moment as we attacked our existential problem. We needed to get an intimate look at what the cleaning mechanism was doing. So, we decided to 3D print a transparent version of the carriage that held the brushes. This would let us view the dirt pickup process directly. We also wanted high-speed video of the cleaning process. We felt we might get the best information if, instead of a blur, we could see the process happening in slow motion. Lacking the proper equipment, we set about to acquire it. And to give us one more perspective, Chris used a band saw to cut the brush carriage and brushes in half making the cross-section visible.

There was a small room built into the high bay (the large, high-ceiling area where we tested robots) at iRobot. I'd often gone there in the past to develop or test aspects of Roomba. In that room, we'd placed a window-sized piece of glass on the topless base of a workbench. Running the robot on the glass allowed one to look upward from underneath. This unique view sometimes revealed how otherwise hidden processes worked. At some point during our flurry of activity I retreated there. My plan was to try to get as good a look as possible at what the cleaning mechanism was actually doing even before the high-speed camera and transparent carriage showed up. The question I asked myself was, "Why doesn't Roomba pick up the dirt?"

Back during the AI Olympics and the Robot Talent show I had formed a mental image of how a carpet sweeper works. Essentially, I imagined that as the horizontal, cylindrical brush spun about its axis, its bristles would pick up particles of dirt when the bristles touched the floor. The brush would rotate around, carrying the dirt with it, until it encountered the teeth of a little comb. The teeth, attached to the dust cup, would comb the dirt from the bristles and it would fall into the dust cup. My imagined process was orderly and elegant with nothing left to chance.

I needed a few more details to implement my Rug Warrior cleaning mechanism. The brush should spin in a direction that opposed the motion of the robot. That is, if I turned on the brush and didn't power the robot's drive wheels, the spinning bush would push the robot backwards. There was a cylindrical enclosure around the brush (like lipstick in a tube). But unlike lipstick, part of the enclosure's side wall was missing so that the bristles could touch the floor. The enclosure was just slightly larger in diameter than the brush.

A force pushed the brush downward such that, as the brush rotated, each bristle came into contact with the floor near the bottom of its swing and slid along the floor for a short distance. If there was a bit of dirt or debris in its path, the bristle picked it up. The enclosure that surrounded the brush then helped the dirt cling to the bristle, or trapped it between bristle and enclosure surface, as the dirt rode up and around to the back side of the enclosure. A rectangular hole at the back of the enclosure gave access to the dust cup. Teeth on the leading edge of the cup protruded through the hole in the enclosure and intersected with the bristles of the brush. Those teeth combed the dirt from the brush as it turned, and the dirt collected in the dust cup.

Now, with the aid of the glass table, I tested my venerable idea. I observed our cleaning mechanism—from the side, from above, and through the glass from below. I seeded the mechanism with different sorts of dirt—from sand to bits of paper to Cheerios and watched what happened. I re-examined other carpet sweepers, comparing their innards with what we'd built. I flicked dirt particles with my finger. My mistake came into focus.

Had I taken the time to look carefully (or maybe consulted a textbook) I could easily have learned the truth long ago. It turned out that my appealing, tidy, and plausible notion of how a carpet sweeper worked—the method I'd built into Rug Warrior, the idea I'd harbored for 13 years, the model on which we'd based Roomba's cleaning mechanism—was utterly and hopelessly wrong. My intuition bore no resemblance to reality.

But, with great relief, at last I did understand. Finally, Melville Bissell could stop spinning in his grave.

So what was really going on? A carpet-sweeper mechanism was built essentially as I described. But it worked in a very different way. Nothing was orderly or elegant, everything was left to chance. As the brush turned, each bristle contacted the floor and, like a little spring, it deflected back a bit. At some point, as it moved forward, the bristle suddenly lost contact with the floor and the energy stored by the bend of the bristle was released, causing it to flick forward. The flicking action did the cleaning. A rapidly spinning brush with hundreds of bristles meant, to paraphrase Jerry Lee Lewis, there was a whole lot of flicking going on.

If a bit of dirt was near the bristle tip when it flicked, then the small particle was propelled violently forward and upward. It might then bounce off the inner wall of the enclosure. Afterwards, it might bounce into the brush and be flung out by centrifugal effects, it might bounce directly back onto the floor, it might bounce from the front of the enclosure to the back and then onto the floor, or (the desired outcome) it might bounce or be flung by the brush through the opening of the enclosure into the dust cup. But the crux of what was occurring inside the enclosure where the brush spun was utter chaos. Particles of dirt were constantly bouncing and being flung in every possible direction.[1]

Only some fraction of the dirt that was flicked forward quickly bounced into the dust cup. Much of it fell back onto the floor. If these falling dirt particles landed in front of the brush or landed within the patch where the bristles were compressed against the floor, they'd get a second chance to be flicked and bounce into the dust cup. But if they happened to land behind the line of contact where the brush touched the floor, the robot would pass them by.

This meant that how far the enclosure wrapped around the brush was a critical parameter of the system that we hadn't appreciated. Ideally the enclosure would surround the brush all the way to the point where the bristles touched the floor—that would prevent any particle from landing behind the brush. In practice, however, the enclosure shouldn't come too near the floor, otherwise any imperfection in the surface (like the nail in our Chicago demo) could snare the robot. This practical detail meant that even an optimally designed carpet-sweeper mechanism allowed a small possibility that a particle of dirt would land behind the brush and not be picked up. So, passing over a spot more than once could help.

The question with which I'd started my analysis—"Why doesn't Roomba pick up the dirt?"—was wrong. In the test that ignited our terror, Roomba actually did pick up the dirt—the bristle brush flicked dirt forward exactly as it should have. But, because the enclosure on the latest prototype didn't wrap quite far enough around the brush, much of that dirt took a ride around the brush and then fell back to the floor in the wrong place—behind the brush's line of contact with the floor.[2] Apparently, we had inadvertently changed the degree to which the enclosure wrapped around the brush from one iteration of the design to the next.

On Winston's next visit to the Roomba office he found everyone smiling broadly. Once we understood how carpet sweepers actually work, we knew what we had to do to fix Roomba.

We called the phenomenon where the brush first picked up a piece of debris and then dropped it on the floor *redeposition*. To eliminate the problem, we extended the enclosure around the brush by a fraction of an inch. Determining the exact amount of extension required some fiddling, but with this done Roomba once again cleaned effectively. We weren't forced to delay commit-to-tooling or gamble on fixing the robot after the molds were cut. Finding the solution had turned out not to take that long; I might even have managed to meet Chris's Wednesday deadline.

NOTES

1 Teeth to comb the brush are not necessary and many carpet sweepers have none. My original sin had been to assume that a minor feature I observed on one of the first carpet sweepers I examined was the enabling mechanism.

2 A Roomba TV commercial aired by iRobot near the time of this writing provided viewers with an instructive perspective looking upward through a transparent floor. The sequence shows the robot encountering what appear to be some spilled oats. Looking carefully, it's possible to observe the brush occasionally pick up an oat, then drop it onto the floor, and then pick it up again I loved watching that commercial.

CHAPTER 16

Robot Replication

FACTORY FINDING

One of the lessons iRobot's association with Hasbro brought home to us was the importance of offshore manufacturing for consumer products. Without that advantage we could see no way to achieve the $200 retail price that Roomba needed. Thus, early on, long before we had a viable design, we began looking for a factory to manufacture Roomba.

To investigate the possibilities, Colin, Winston, and Phil took the nominally 20-hour air journey from Boston to Hong Kong. From there they set forth into southern China, visiting possible manufacturers. They stopped for a day each at four factories that looked like good possibilities.[1]

One of the factories was too big. It seemed that Roomba's anticipated production run would include too few units to matter much to them, meaning that our product might get lost in the shuffle. A second factory was very appealing in several ways, and they were eager for the business. But they were small and hadn't built anything nearly as complex as Roomba. The robot would be a big stretch for them.

The third factory stood out for unsettling reasons. They proudly showed Colin, Winston, and Phil the smooth functioning of their backroom operation—where they systematically reverse engineered products others had built for the purpose of knocking them off. That factory was quickly dropped from the list.

The remaining factory, called Jetta, seemed to occupy the sweet spot.

Jetta manufactured toys and had been doing so for some 25 years. Remember the simple toys that always accompanied McDonald's Happy Meals? Jetta was one of the companies that made them. But there was a pecking order among factories in China, and manufacturers of simple toys did not occupy the most prestigious rung. The ambitious folks at Jetta wanted to improve their standing.

The business had been founded by engineers, and they naturally wished to manufacture more sophisticated products. Landing the contract to build a consumer floor-cleaning robot seemed just the ticket. Jetta aggressively pursued a deal with iRobot, and in the end an agreement was struck. Roombas would be manufactured at Jetta's facility in Panyu, China.

SIDE BRUSH

Many elements of the robot competed with each other for space. Nowhere was the competition fiercer than on the robot's underside. We wanted to install a cleaning mechanism that spanned the full width of the robot, so that the sides of the main brush were flush with the sides of the robot. This would enable the robot to clean the floor all the way to the edge of the baseboard.

Unfortunately, the robot's drive wheels also wanted to be nearly flush with the sides of the robot. This meant that when driving along a wall, the edge of the brush would come no closer to that wall than about three inches.

Accepting that gap (as did the designers of Trilobite and most other floor-cleaning robot attempts) meant that every floor would be incompletely cleaned—a dirty strip would border every wall and object in the room. Also, because of the robot's round shape, the main brush couldn't reach into the corner. The robot was round, the corner square, and the difference was a small area that would never be cleaned. Over time, more and more dirt would accumulate along the edges of walls and in the corners, until the customer felt compelled to remove it manually.

We first looked for a synergy with the vacuum we'd had to install. The vacuum sucked air in through the narrow inlet on the robot's bottom, but after the dirt had been filtered into the collection chamber, the clean exhaust had to go somewhere. Was there some clever way to duct the air such that it would blow dirt away from the wall edge or out of corners and into the path of the main brush? We did some initial experiments, but they just produced a mess. The directed air blew dust everywhere, scattering the dirt rather than collecting it.

Next, we tried adding another brush to accomplish our purpose. We called the new feature the *side brush*. Unlike the main brush, with its horizontal axis of rotation, the side brush, like the Bugs Bunny's Tasmanian

Devil, would whirl about a vertical center line. The side brush would be mounted ahead of the drive wheel on one side of the robot, with the bristles of the brush extending beyond the footprint of the robot so as to reach into corners.

The idea seemed simple enough. But execution gave rise to two problems. One was unavoidable, the other we created for ourselves.

The unavoidable problem related to autonomy. We had gone to great lengths to maximize the robot's mobility—its capacity to run over all manner of impediments and just keep going. The springs that forced the wheel modules down helped with that, as did the floating carriage that housed the main brush. But now we were adding one more item that would touch and therefore apply force to the floor. That gave objects on the floor an opportunity to apply force to the robot. Would the side brush harm mobility, making the robot more likely to get stuck?

The fact that we couldn't accept a decrease in mobility created an awkward requirement for the brush. In favorable situations it must help (enhance cleaning), in unfavorable situations it must do no harm (preserve mobility). An example of a favorable situation is driving along a tiled floor next to the wall. Here the side brush must be very effective at moving dirt into the path of the main brush. An unfavorable example is when the robot tries to drive up onto a braided rug causing the rug to invade the side brush's space. The side brush must not generate forces that will prevent the robot from surmounting the rug.

The first side brush design we tried, with a gearbox designed by another iRobot engineer,[2] had many bristles radiating from a central hub—like an extremely squat Christmas tree. We tilted the hub just slightly, so that the bristles only touched the floor outside the footprint of the robot. At the point where the spinning bristles were within the path of the main brush they were no longer in contact with the floor, thus releasing the dirt from the side brush.

This arrangement worked fine in favorable situations. Under unfavorable conditions the bristles tied themselves in knots. The knots were permanent. We tried fewer bristles, we tried bundling the bristles, we tried bristles made of different materials. We got knots, knots, and more knots.

Then we tried attaching rubber arms to the hubs and adding bristles to the end. Knots developed less frequently but the side brush sometimes hurt the robot's mobility. The trick seemed to relate to the exact shape and springiness of the arms—they had to taper in just the right way and be made of the right material.

Finding the sweet spot was a frustrating process of trial and error that went right down to the wire. Actually it went over the wire. At the time we pulled the commit-to-tooling trigger, and the factory began cutting the molds, we had not yet solved the side brush problem. But the factory graciously accommodated us by agreeing to carve the side brush into one of the last molds they made rather than the first. That bought us a couple of precious weeks; finally we converged on a design that worked, and the molds could be machined.

The side-brush problem we created for ourselves resulted from our extreme frugality. The robot had four motors at this point: one each for the two drive motors, one for the main brush, and one for the vacuum. The side brush needed to spin, but we were very reluctant to buy a fifth motor, and the electronics required to drive it.

The main-brush motor had sufficient power to run the side brush as well. The problem was that it spun on a horizontal rather than a vertical axis and the output of the motor went through a gear box that wouldn't stay put—it had to move up and down with the brush carriage.

Over numerous iterations, Eliot fought the good fight. He sought a clever way to actuate the side brush using the main brush's motor. He experimented with sliding, contoured worm gears and other exotic mechanicals. But in the end, the challenge was a linkage too far. So, we gave up and bought motor number five.

BITS BEFORE BOARDS

Usually when building a robot or other sophisticated device, software is the last item to be checked off. Engineers count on being able to easily change the software right up to the last minute. That way, if anything went wrong late in development, they would be able to "fix it in software." The process was to first devise some clever workaround for the mechanical or electrical issue that caused the trouble and then plug a cable into the product and download new code to fix (or more typically, mitigate) the problem. But circumstances prevented us from doing that with Roomba.

Roomba's program would be stored within our microprocessor's on board, read-only memory, where the program code would be "burned in" to the chip at the chip manufacturer's factory. After programming, the chips would be shipped to Jetta, where they'd be soldered into Roomba's circuit board. The circuit board could then be installed in the robot. Once the chips were programmed at the factory, changing Roomba's program was no longer possible.

Logistics were such that Phil had to finish his program before we gave the final OK to the factory declaring that all the files describing Roomba were complete. After shipping the code off to the chip manufacturer but before we gave the final OK to the factory, we found a new problem.

As the side brush turned, it swept out a disk-shaped volume. One of our cliff sensors peered through that volume. Such brush/sensor interference was less than ideal, but we accepted the compromise so that everything would fit in its proper place under the robot. After all, the arms of the side brush were narrow and their brief, periodic interruptions of the cliff sensor's view shouldn't have caused a problem. Unfortunately it did, sometimes. We discovered that some robots, operating on some types of floors, stuttered. In unfavorable circumstances, the robot interpreted the side brush passing by as the floor vanishing and then reappearing—that made Roomba repeatedly start and then curtail its cliff-avoidance behavior.

The obvious solution was to tweak the code. We might have told the robot not to react unless the floor remained missing for some number of milliseconds. But we couldn't do that, the code was finalized, and the microprocessor manufacturer had started programming our chips.

The only solution was to take the uncommon step of fixing a late-stage problem in hardware. Since we were still able to change Roomba's printed circuit board, Chris added components that filtered out the regular interruptions of one cliff sensor by the side brush and Roomba regained its cleaning fluency.

ELIOT IN THE MIDDLE KINGDOM

The terminal phase of the Roomba project was frenetic and trying for us all, but probably Eliot felt the greatest impact. We did as much testing as we could in as many different homes as we were able, and we frequently uncovered circumstances that challenged Roomba's autonomy; i.e., we found situations and arrangements of obstacles that caused the robot to become stuck or nearly so. When this happened, we would first figure out why and then make improvements by tweaking the design—changing either the programming or the mechanics, or both. (The electronics were mostly stable by this point, so Chris switched his focus from electrical engineering to being our test engineer.) As described earlier, because of our lack of siloed modularity, changes to Roomba's mechanical design tended to reverberate—a change in one part necessitated changes in others. This created copious CAD and analytic work for Eliot.

The project could have made good use of one or two additional mechanical engineers to handle the load. But as it was, most of the work could only be done by Eliot and only a dwindling supply of time was left to complete it.

Eliot worked like a man possessed. Had an angry animatronic buffalo been hot on his heels, he could have toiled neither harder nor longer. Underlying the stress was the urgency of the impending deadlines for a fall delivery to our launch partners. "If we miss them," Eliot thought, "we won't get another chance."

And so day after day and frequently into the night Eliot worked. At his computer, he tweaked the model to fix the problem of the day, chasing down all the ripples those changes gave rise to. At least, Eliot comforted himself, the torture would end as soon as he handed off the files to the factory. Telling the factory "Go!" would be his key to freedom.

Finally, the glorious day arrived, and Eliot sent his 100 percent complete files to China. But just to make sure things got off to a good start, Eliot was pressed upon to make a one- or two-week trip to the East Asia. On site at the factory, he would be able to soothe any growing pains that might develop as workers there attempted to convert his CAD files into a real product.

It turned out that not every piece of essential information was encoded into those files, and that certain realities of manufacturing required subtle changes to Roomba's design. That was to be expected as we had never done a production handoff on our own before, and Jetta had never before built a product this complex.

Some of the first parts the factory made didn't work. The drive train for the main brush was off; gears either didn't mesh or interfered with each other, and the brush itself had issues. It wasn't nearly as effective at picking up dirt as the Bissell brushes we'd long used to prototype Roomba. But that was just the tip of the iceberg; there turned out to be many issues preventing a smooth transition to manufacturing.

With acute dismay, Eliot watched as the shackles that bound him to an interminable workload snapped shut around his ankles once again. It wasn't just a few early-stage difficulties that he needed to resolve, there turned out to be a weary abundance of subtle issues and problems to untangle in order to transform computer files into robots. Two weeks in Panyu turned into six months. Mercifully, there were several interim trips home.

Initially, Eliot and his counterparts at the factory struggled to work together effectively. The staff found the product complex and unfamiliar, different groups had different areas of responsibility for its construction, and Eliot spoke no Chinese. As technical problems cropped up Eliot would discuss changes with the few resident English speakers, make drawings, and write instructions showing how a fix could be effected. But then nothing happened. He began to wonder if he was failing to see past some unspoken cultural impediment.

At some point Eliot realized that Jetta's engineers were fluent in the same Pro/Engineer (Pro/E) CAD system he'd used to model all of Roomba's parts. Discovering that common language made all the difference. Rather than trying to describe the changes he wanted others to make, Eliot could simply tweak his original files using the factory's CAD system. This done, workers knew exactly how to proceed—Eliot's changes would become real in mere hours. The transformation was magical; it was as if a genie spoken to in Pro/E would immediately grant wishes cut in steel.

Because robot vacuums were completely new, no specific standards existed. But the FCC raised its hand—they had a standard that did apply. Every product must refrain from emitting too much electrical interference. Cheap electric motors like Roomba's make tiny sparks as they spin. Sparks are notorious for creating electrical interference, and the electrical noise our vacuum motor hissed exceeded the FCC's threshold.[3] We needed to add a circuit board to the back of the vacuum motor to suppress the noise. Although tiny, this new circuit board would bump into the plastic surrounding the vacuum, so that part of Roomba had to be changed. This would be only the first of many such tweaks.

As changes were made to the molds, new sets of parts were produced. The parts were assembled into robots and the robots were tested. In one such test the newly assembled robot's cleaning head suddenly ceased functioning. This was perplexing because the cleaning head's plastic parts hadn't been altered since the previous successful test. Why would it fail now?

A careful examination revealed how the head had failed: part of it melted. Not *every* mechanical part of the robot was made of plastic. The stubby shafts on which gears spun were steel. One of the shafts on the failing robot had gotten so hot that it melted its way right out of the gear box. Why had this shaft succumbed to a high fever while earlier ones kept their cool?

Steel gear shafts start out as long rods. The rods are then cut into short pieces with the correct length to fit the gear box. Previously, workers had cut the shafts to length using a hacksaw. But then they realized things would go much faster if the cutting were done with *powered shears*. Powered shears work like kitchen scissors on steroids.

Unfortunately, the shears delicately deformed the shafts. During operation, when the gear spun rapidly, the slightly flattened shaft created friction that heated the steel to the melting point of plastic. To the naked eye, the shear-cut shafts appeared perfectly round. Eliot was able to identify the flaw only by measuring the shaft with a *micrometer*.[4] A new way was found to make the gear shafts that preserved their shape.

On another day another test uncovered another problem. Some newly assembled Roombas were sent to clean a deep-pile carpet. The cleaning heads immediately ground to a halt. To the eye, the salient parts of the brushes Jetta built into the cleaning mechanism looked identical to the familiar Bissell brushes. But once again, Eliot's micrometer found the trouble. The bristles of the Jetta-built brushes were too fat by five-one-thousandths of an inch. That made the bristles push too hard against the carpet stopping the brush from turning. Substituting bristles with a smaller diameter fixed the problem.

These episodes brought home the fact that in mass production everything counts, even subtle things no one can see.

Another issue that cropped up was the rubber material used to form the tread of Roomba's tires. The material we'd used for 3D printing worked well, providing all the traction the robot needed. But that same substance wasn't appropriate for injection molding and the standard material Jetta tried made treads that had too little friction—allowing Roomba to slide and spin out. The issue set in motion a month-long process to find a rubber compound that would work.

The near 12-hour time difference between Panyu and Somerville produced a 24-hour call and response cycle. At the end of most of his days, Eliot would email a list of problems and difficulties he was encountering at the factory. We received the list in the morning Eastern time, worked on them all day, and tried to send off solutions by the end of our day. Eliot received these as his day began, and so on.

Over the spring into the summer of 2002 the wrinkles in making the robot's parts were gradually ironed out. The factory began to plan how they would implement the production of thousands of Roombas. The first phase involved Eliot sitting in a room with a half dozen Jetta line supervisors, thoroughly reviewing the assembly process. Afterward, the supervisors would train the workers who would perform the actual construction.

It was all very cerebral. Then, after a two-week trip home, Eliot returned to China to find things had gotten serious during his absence. Now there was an enormous room packed with people all busily engaged in bringing Roomba to life. In the best tradition of Henry Ford, managers had set up an intricate assembly line with table after table arranged end-to-end snaking around the factory floor. Powered screwdrivers on retractable tethers dangled from above ready for workers to grab and attach the many fasteners Roomba required. There were soldering irons, testing areas, and a zone where packaged robots were stockpiled for shipment. Jetta's Roomba manufacturing area matched the size of iRobot's head-quarters and employed even more people. A robot-building juggernaut was revving up.

All that he saw—the people, the equipment, the activity—traced back to the CAD files Eliot had created and the assembly instructions he'd pro-vided. The responsibility was daunting. Eliot crossed his fingers, "Hope I didn't forget anything," he thought.

One of the trickiest items in the entire assembly turned out to be that little piece of string of which we were so proud, the one that let the main brush dynamically adjust itself to apply just the right force to the floor. A knot needed to be tied in the string at a particular spot. Workers found it difficult to position the knot correctly and the time this took to get it right produced a bottleneck. Eliot offered a free dinner to anyone who could come up with a better way to implement that step, but no one ever claimed the prize.

Remaining kinks continued to be worked out of the assembly process. The multi-page things-to-fix lists that went back and forth between China and Somerville every morning and night became shorter and shorter. The assembly lines hummed and rooms full of Roombas in their little testing boxes (each a replica of the original carpet and tile combo created in Somerville) whirred along.

It all came together just in time—the deadline for shipping the first ten thousand robots to our initial sales outlets loomed. The least costly way to get products from southern China to the US was via a slow container ship. When we began the manufacturing process we planned to use that method. But as problems were discovered and gradually solved, we missed the deadline—that form of transit took several weeks. For somewhat more money Roombas could ride a faster ship. We missed that deadline too. By the time the factory was ready with enough robots to supply our main vendors there was only one option left. Roombas had to fly.

The shipping agent approached Eliot with a form. Sign the paper and Roombas would take wing for America. Thinking, "Well, I hope this works," Eliot signed the form.

It was almost exactly two years earlier that Eliot had first seen the Scamp prototype. He'd observed the robot's simple sensors, straightforward mechanical design, and few motors. "How hard could that be?" he'd thought then. Now he knew.

NOTES

1 Finding the most appropriate factories among the literally thousands of possibilities was no easy feat. Our search was facilitated by Chris Dooley, iRobot's Director of Consumer Operations, whose experiences before coming to the company had granted him expertise in this area.

2 By this point, late in the project, Eliot was overwhelmed by an avalanche of last-minute mechanical changes. We gratefully accepted help from a few folks outside the Roomba core team. One was mechanical engineer extraordinaire Steve Hickey who designed a small, efficient gearbox for Roomba's side brush.

3 The very first practical radio system built by Guglielmo Marconi in the 1890s was based on a spark-gap transmitter. Marconi used high-powered sparks to generate electromagnetic waves that, unlike now, were not tuned to a particular frequency. Indeed, Marconi's equipment consumed the *entire* radio spectrum in order to transmit just a single message—no one else could have broadcast while Marconi was on the air. This ability of sparks to interfere across a broad range of frequencies requires that they be strongly suppressed at the source.

4 A micrometer is a gauge that can measure thickness with very high precision.

CHAPTER 17

Shock and Awfulness

WINSTON TAO MANAGED THE Roomba project. But his "project manager" title didn't begin to describe all Winston did for the team. He was our enabler. His interest and assistance had gotten the project off the ground. He was our champion. His tactics insulated us from the buffeting of company politics. He was the fierce and savvy defender of our team.

But managing the interface between team and company was the lesser of Winston's contributions to the project. Much more important was a skill he shared with few corporate managers—the ability to effectively direct highly innovative teammates. Exactly that skill was critical in order to bring into alignment two often-at-odds imperatives—the internal motivation of creative people and the needs of the business.

This always required patience, honesty, space, and respect for our team's creative process. Sometimes Winston (often with Phil's assistance) had to manage indirectly. When we were unwilling or unable to see how some technical matter would affect customers' perception of the robot, Winston found creative ways to show us. His enlightened approach meant that we took ownership of any necessary changes to the robot (like the late addition of a vacuum) rather than resenting having changes imposed on us (like SC Johnson's dust cup consumable).

Winston's stated aim was to build a self-managing team. He did. Each of us stepped up from time to time—as needs required and expertise enabled—to lead some aspect of the development. This, among other positives, made working on the Roomba team the most fun and fulfilling experience of my professional career. My feelings were not unique.

DOI: 10.1201/9781003540489-17

In early May 2002 we achieved a momentous milestone with the release of the CAD files that defined Roomba to the factory. Although launch was still a few months away, file-release (minus some lingering details) marked the end of the development phase of the Roomba project.

Roughly a week later, our happy accomplishment was overshadowed by a baffling and disheartening event. Our champion was exiled. With the suddenness of an exploding battery pack, Winston and a few other employees were let go. The team had no inkling any such thing was being considered, we were stunned and upset. The CEO called us in and recited a rationale for the actions taken.

Winston had unfailingly kept the Roomba team insulated from the intrigues of office politics. Regrettably, he was unable to erect similar defenses for himself. The next political strafing run would find the team's shields at zero percent.

CHAPTER 18

Pepsi to Pop

LAUNCH

On Wednesday, September 18, 2002, an iRobot press release announced that Roomba was available for sale.[1] Simultaneously, several prearranged news stories broke. A few days later another event in the marketing master plan took place. A type of vehicle I'd previously seen only from afar pulled into the alley behind the Twin City Plaza. The vehicle was a satellite truck. On its top were parabolic dish antennas carefully aimed upward and southward toward precise spots 23,000 miles above the equator. That's where a constellation of communications satellites sat motionless as if bolted to the ancient's concept of the starry firmament. Traveling at 1.91 miles per second, a *geostationary* satellite completes one orbit in 24 hours. Thus each appears to hover motionless above a spot on the earth.

I walked outside to take a closer look at the truck. Peeking through the open door I was awestruck by the sheer density of the exotic equipment filling every square inch of rack space inside. Curiosity overcame my inner introvert, and I actually initiated a conversation with the engineers monitoring the equipment. They told me which satellites the antennas were pointed at. They described uplink and downlink capabilities. As they spoke, they pointed at oscilloscopes displaying signal strength, at TV monitors showing live feeds and test patterns, and they gestured toward lights and dials and all the rest. If only my ten-year-old self could have witnessed such wonders! The truck that came to Somerville that day held technology light years beyond the crude apparatus I found so inspiring in the dusty TV repair shop of my youth.

DOI: 10.1201/9781003540489-18

Roomba's launch meant that for the first time consumers could purchase an affordable robot to clean their floors. The satellite truck was there to help iRobot so inform the world. A thick, black cable snaked from the truck, up through the stairwell, along a hallway to a console in the high bay. Connected to the console were cameras and audio equipment. These were directed toward a faux living room we had constructed in the high bay. Our CEO sat on a four-legged stool in that space, Roomba in hand, waiting to be interviewed. The iRobot marketing department had arranged with a dozen or more local TV stations across the country to do a piece on Roomba during their midday news programs. The satellite truck enabled the needed connections.

So, again and again, as noontime marched westward across the country, local anchors in different cities would introduce Roomba and our CEO to their viewers and ask the same questions. What is Roomba? How does it work? How much does it cost? When will it be available? Many hosts seemed a bit incredulous that the answer to the last question was, "Now!"

Wonder filled that day. After a torturous, multi-year gestation, our mind-child was born, and the world was eager to hear the news. Every precious second of that time was exhilarating and joyous. Every interview given, every news story published or broadcast, and every mention on the internet was an immense delight. "Savor the moment," I told my colleagues, "It will never be better than this."

On that day, at least emotionally, it was impossible to imagine that Roomba could fail. (This was testament to the insufficiency of my imagination.) The team, now empty-nesters all, had done everything we could do to give our mind-child the best possible chance. Roomba was now on its own in the world and would stand or fall in response to forces we mind-parents could little influence, let alone control. Marketing, business strategy, random chance, and fickle consumer preference would decide Roomba's fate. And Roomba, despite our best efforts, had at least one potentially fatal foible. In contrast to Electrolux's highly refined Trilobite, Roomba was more adolescent than adult.

The marketing team had done a superb job as our initial promotional efforts could hardly have been more buzzworthy. In addition to TV and radio interviews, an article titled "Maid to Order"[2] appeared in *Time* magazine. The reviewer, Lev Grossman, spoke of Roomba in glowing terms. Other prestigious publications including the *Wall Street Journal*,[3] *Technology Review*,[4] and *Forbes*[5] also heralded the news as did many smaller concerns.

Immediately after launch we began scouring the internet for mentions of Roomba. Is anyone talking about it? Reviewing it? Complaining about it?

The first few weeks were unsatisfying. Google searches yielded only references to news stories and mentions we already knew about. But our investigations did uncover one unexpected item. There turned out to be a Cuban restaurant in New Haven, Connecticut named "Roomba." I always wondered what effect the robot's popularity had on them.

Gradually, more and more search hits began to appear. Additional news stories were reported, more reviews were composed, and individuals and thought leaders wrote about Roomba.

For many weeks we eagerly anticipated a review from *Consumer Reports*, the last word in appliance evaluation. The Roomba team were all fans of the magazine, so our disappointment was acute when CR finally published their results. The author gave Roomba demerits because it took longer to clean a room than a person would. "You don't get it! It doesn't matter how long the robot takes because you don't have to do it," we screamed in the direction of Yonkers, New York.

We had delivered a Roomba to the influential technology writer Walt Mossberg for the first round of press coverage. Later he told us that he'd let his secretary use the robot in her office one day. She set the robot cleaning while she ran a quick errand. Upon her return she found that Roomba had pushed her office door closed, locking her out. But Walt gave us a good review anyway.

People began buying Roombas in large numbers. Initially it was available only from our website and the specialty retailers Brookstone, The Sharper Image, and Hammacher Schlemmer.[6] Even so, we quickly sold the initial production run of 25,000 Roombas we had ordered from the factory, thus zooming past the HeathKit Hero, Paul's and my robot-sales role model.

An online Roomba users' forum spontaneously arose to serve our new customers. We'd hoped something of that nature would happen. Unfortunately, many of the messages exchanged related to how to fix the various Roomba problems that quickly emerged.

STUFF PEOPLE LIKED

Roomba worked. But even better than that, customers could *see* that Roomba worked. Run Roomba on any floor—even one that appears clean—and then examine the dust cup. You would find it filled with debris.

A manual vacuum offered less conclusive, less frequent evidence of its proper operation. You heard its powerful motor, you could see it gobble up a few bits of obvious dirt, eventually you changed the dust bag—but on any single run you had to take it on faith that the machine was doing a good job. In contrast Roomba demonstrated its utility every time you used it. In the early days it was always a shock, after running Roomba, to see how much dirt it found on a floor that looked clean to begin with. That contributed to a satisfying user experience.

The dust cup started fairly small, and it got even smaller when we had to cram a vacuum into that space. But what seemed to us an embarrassingly petit collection volume probably turned out to be a marketing asset. The fact that users needed to dump the collected detritus after every use meant that they got instant feedback validating the robot's efficacy. Demonstrating the robot to their friends, customers could say, "Look at all the dirt Roomba picked up that I didn't even know was there!"

Before launch we had worried greatly about what to call Roomba. Dare we describe it as a robot? Or might movies like *The Terminator* have poisoned the well, making customers fearful of having a possibly untrustworthy automaton running amok inside their homes? To err on the side of caution, we decided to avoid the term. On the box, Roomba was described only as an "Intelligent FloorVac." Nowhere did we say it was a robot. But our caution turned out to be unmerited. Customers seemed to feel no hesitancy toward giving Roomba free run of their homes. Comments to our customer support representatives suggested that owners *liked* the fact that it was a robot. So, subsequent models expressed their robot pride right on the box.

From the earliest days of the project, we technical team members firmly and unanimously agreed that we would eschew cuteness. Roomba would have no googly eyes, no face, no contrived anthropomorphism of any sort. We were building a tool, period. Yet Roomba developed a personality both in spite of, and because of us.

The way Roomba moved mystified and intrigued users. Although its motions were random, owners more often than not ascribed calculated intent to its wanderings. Had it performed an efficient boustrophedon pattern (that we would certainly have implemented had we been able) Roomba would likely have seemed more robot-like and less interesting. When Roomba followed walls its characteristic waggling motions also made it seem more organic and alive. Roomba's tenacity was another appealing quality. When it worked its way into a tight spot, it just kept trying until it (usually) got out. The triumphant little tune it played upon finishing a cleaning run made Roomba seem proud of the job it had done. And the "uh-oh" it emitted to request help when something went wrong was more endearing than annoying.

Cleaning my house with Roomba felt like a partnership. While I dusted or decluttered, Roomba did its part by cleaning the floor. I liked that. I suspect others did too.

Many owners named their Roombas. They chose monikers like: Rosie, Wall-E, Optimus Grime, Meryl Sweep, Marvin, Jeeves, Alfred, and Alice. A study[7] from a few years later found that 80 percent or more of users named their robot as they would a pet. Although the Roomba team had aimed only to build a useful tool, early anecdotes about owners naming their robots let us know we had done something more than that. After all, how often do people bestow such honor on their other tools? Have you ever named, say, your electric toothbrush?

GOLIATH'S VERDICT

Electrolux became aware of Roomba very soon after we launched. What cunning disguise or clever ploy Electrolux's covert agents used to arrange an acquisition we never learned. Maybe they just ordered our robot from Hammacher Schlemmer. (How dull!) But in any case, Roombas quickly arrived in Västervik and Stockholm.

The engineers examined our mind-child with suspicion. They powered up Roomba and watched it work. It seemed to the Electrolux folk that what we had built wasn't really a vacuum cleaner.[8] A true vacuum couldn't be built for the price we charged. And, in their estimation, it didn't pick up dirt as well as their machine. But they took our work seriously and, although Roomba wasn't beautiful and didn't look to be of the same quality as Trilobite, they decided to subject it to a definitive life test. Could plain, cheap, feeble Roomba go the distance?

The results did not impress the Electrolux engineers. They confirmed that Roomba's quality indeed did not match Electrolux's high production standards. Likely some thought, "Roomba is just a passing fad. Trilobite is better."

Well, one out of two.

I know how the Electrolux engineers must have felt because I've been in that spot myself—that irksome place where customers just don't get it. Where, by every metric you believe important, your product is superior. And yet, uncaring customers insist on buying your competitor's inferior item, leaving yours to sit unloved on store shelves. Then later, the store evicts your darling from even that lonely perch. As tough as real-world robots must be those of us who design them need even thicker skins.[9]

UH-OH

Almost immediately, Roombas began to break.

As noted, none of the Roomba core-team members had ever shipped a consumer product before. Eliot came the closest, with the electronic fishing game he designed at his first job. But even he had never before seen all the stages of a product—from idea through design to manufacturing, and then distribution and customer service.

We knew that reliability testing was important, and we gave it our best effort. For example, Paul built a mechanism that tested drive-motor life by attaching a heavy wheel-sized weight to the motor output and then programming a controller to drive the wheel back and forth for many hours to simulate the wear and tear a robot would endure during its lifetime. We set up stacked pens where we ran robots continuously, day and night. And we did many other tests as well. But, as the Electrolux engineers proved with their life test, we didn't do enough.

Operating in the real world, Roombas found more ways to fail than we had imagined or discovered in the lab. Despite our tests, some wheel motors burned out too soon. The rubber treads that gave Roomba traction on slippery floors occasionally tore and fell off. On their own, some users came up with ways to glue the treads back on, sharing their fixes on the forum. Hair wrapped around the axles of the main brush. Left too long this accumulated until the bulge of hair pushed against the assembly, producing friction and heat. In the worst case, this caused the gear box that connected the motor, bristle brush, and flapper brush to melt.

It hadn't occurred to us that the belts connecting drive motors to wheels would shed little particles of rubber; over time, these particles sometimes clogged the sensors that constituted Roomba's wheel encoders. The blinded sensors then couldn't tell whether the robot's wheels were turning. This condition caused Roomba to perform a characteristic but unproductive little back and forth dance.

iRobot made good on robots that went bad. When users had a problem with Roomba, the first step was to call our customer support line. During the first few weeks the robot was available, all members of the development team were required to enter the phone-answering rotation. I thought this was a great plan because it connected us developers to real customers, letting us learn firsthand about any problems the robot might have. (During his stints on the phone lines Eliot would visibly flinch when early users asked about removing the brushes.)

This would help us to rapidly improve the product. At the same time, I felt speaking with customers was a horrible plan because the anticipation of talking to real people initiated heart-pounding, gut-wrenching episodes for an introvert like me. Unbidden, my mind conjured scenarios of customers screaming accusations of false advertising and fraud. But reality turned out to be a friendlier place than my imagination. The people I spoke with were all understanding and eager to try my suggestions to get their robot working again.

When a phone call couldn't resolve the problem, we'd take the robot back and send out a new one. For some customers this process happened multiple times. Once, while Chris was the operator on duty, a call came in from a most patient gentleman. Somewhat wistfully the man explained that his robots had had a series of failures, the fourth one had just died, and would we please send him a fifth?

It's nearly impossible to imagine that a customer who had, say, four toasters of the same make and model fail would express any interest in a fifth. More likely, they'd indignantly demand their money back forthwith. Somehow, Roomba brought out a more nurturing side in many of our customers.

A request that arrived not infrequently was that we repair rather than replace an afflicted Roomba. Sometimes the customer would explain that he, she, or the kids had become attached to the little guy, and they simply must have their ailing servant restored to health and returned to them. That was another response the team didn't anticipate.

Other "customer" responses were more humorous. Jeff Ostaszewski had handled more than his share of the initial customer calls, as iRobot initially had nothing resembling a customer support center. He began to look grizzled and worn due to the long hours but gracefully responded to the multitude of user questions. In one case, since the "customer" had an intense technical question about the mechanical components, he handed the phone to Eliot.

Eliot then heard a very unusual "user question" couched in a heavy accent. "On zee package, it states zee Roomba cleaning uses 15 watts." Eliot thought for a second, mentally multiplying the nearly 14-volt levels of the battery pack at full charge by the brush motor's draw of about an amp.

"Yup," he said into the phone. "That's about right." Eliot then distinctly heard a different voice yell, "That is impossible!" as if a second person were listening in on the line. The original voice said, with a touch of weariness, "Zank you very much," then hung up.

Apparently, other engineering teams regarded the efficiency we had achieved with Roomba to be an unattainable goal!

DAVE CHAPPELLE SAVES ROOMBA

A young man saunters about the kitchen of his date's apartment while she prepares for their evening out. He's surprised when he sees a small robot vacuum cleaner rolling down the hall toward him. He tosses it a chip from a bowl on the counter and the robot promptly gobbles it up. Next, he waves the soda he's drinking at the robot prompting it to come closer. The vacuum noise intensifies as the robot bumps his feet. Concerned, the young man moves rapidly away only to fall onto the floor propped up by the couch. The vacuum immediately ingests his trousers.

Hearing the commotion, his date comes to investigate. On his feet once again, still sipping his soda he says, "Your vacuum cleaner ate my pants. Wasn't nothing I could do." His date rolls her eyes and walks away.

The above describes a 30-second spot produced by PepsiCo Inc., the makers of the soft drink Pepsi. Starring Dave Chappelle as the hapless boyfriend, the ad hit the airwaves around Thanksgiving 2003. It changed Roomba's fortunes.

Roomba's first year of sales starting in 2002 had amazed us all. The second season iRobot decided to stock up for success—only to find, late in the year, that thousands of robots were sitting on warehouse shelves unsold.[10] Then Pepsi broadcast their commercial and sales immediately accelerated.

I profess innocence of all things marketing. How the Chappelle commercial could sell Pepsi is unfathomable to me. How it sold Roombas, doubly so. (To the trained eye, the robot in the commercial neither looks nor behaves like our robot.) But perhaps Roomba benefited because the commercial suggested to tens of millions of viewers that robot vacuum cleaners are here and now. In any case it seems to have been one more piece of providence for a robot that had more than its share of good luck—at least in its later years.

PET MOBILITY

I don't recall it occurring to anyone on the Roomba team that owners would use their robots for purposes other than cleaning floors. Perhaps we lacked imagination. Roomba owners were afflicted with no such deficit;

they put cats on top of Roomba and filmed them riding around. Then they put those cats in costumes. They transported small dogs, chickens, guinea pigs, bunnies, hamsters, opossums, and even babies.

They taped balloons to one Roomba, attached a knife to another, and set them both running—not recommended! The most ill-advised usage I've seen to date was casting Roomba as the protagonist in a real-life game of *Frogger*. Yes, at least one owner aimed their robot across a busy street and let it go. Please don't do that.

Roomba was cast in numerous TV shows. Perhaps the robot's most unsettling appearance was on the show *Arrested Development*. There Buster Bluth (Tony Hale), having become part machine himself (when his hand was replaced with a hook following an unfortunate shark encounter), becomes enamored with Roomba. Fortunately, any unspeakable acts that may have occurred happened off screen.

Roomba helped mystery show sleuths solve crimes at least once by collecting crucial evidence while cleaning. Roomba also helped the TV good guys evade the bad, in one case by being engineered to create a distraction by falling down the stairs. (Hollywood tinkering enabled this event, as Roomba would refuse such a stunt in normal practice.)

CHRISTMAS PRESENT

A year or two after Roomba's launch iRobot held their annual holiday party on a boat in Boston harbor. Along with most of the company, I attended with my wife and two young daughters. It was a festive occasion, spirits were buoyed by the twin successes of Roomba and iRobot's other robot, PackBot (the teleoperated machine proved popular with the military).

Joining the celebrations that night was a woman I didn't know. She walked with assistance and although she appeared quite happy to be there, she seemed a little out of place. Midway through the evening our revelry paused so we could hear her story. The woman, whom we'll call Doris, had been invited to our holiday party because of the touching account she'd given one of iRobot's customer service representatives. A few years earlier Doris had been stricken with a condition that affected her mobility. In the before time, she had socialized frequently, often inviting friends to her home for dinners and visits. But since becoming ill, she'd mostly stopped seeing people—that left her feeling isolated and sad.

The reason she stopped seeing friends was not because her mobility issues prevented it. The problem was embarrassment. Doris was embarrassed that her house was dirty, and she couldn't clean it. Her illness made it physically painful to operate her standard vacuum cleaner. So she mostly didn't, and dirt accumulated.

Then Doris discovered Roomba. It made a huge difference in her life. While she couldn't push around a regular vacuum cleaner anymore, she could easily push Roomba's power button and empty the dirt container when it finished cleaning. Roomba eliminated her embarrassment along with the dirt, enabling her to reengage with her friends. Doris thanked us profusely for the wonderful product we'd built. Her story left many of us in tears—the memory still has that effect on me.

I'd wanted to build a little floor-cleaning robot because robots are fun, and it seemed like the time was ripe for such a machine. But until Doris told us her story that evening, it never occurred to me how profoundly my work could touch another person's life.

INDUCTION

It happened two or three Saturdays in a row in November 1975. After hours of studying physics, I'd emerge from my room late in the evening to find the floor of my graduate school dorm deserted. Where was everyone? It must be just a coincidence that all my neighbors were out at the same time, I figured. Then one Saturday I happened to wander down to the TV lounge. There, packed in like proverbial sardines, laughing uproariously were all my dormmates. They were watching a new show I'd never heard of called *Saturday Night Live*. I squeezed into the room, claimed a rare speck of unoccupied real estate, and became an instant fan. I've been watching *SNL* ever since.

Twenty-nine years later, on December 18, 2004, to be precise, I was engaged in my usual Saturday night TV viewing passion when a sketch began. The camera zoomed in on a genteel woman lounging on a sofa. She tells us that feeling fresh makes her confident. Next, we see a woman preparing for bed, as she says that she's at her best when she feels clean. A kitchen scene follows where a woman is kneading bread. Her hectic life gives her no time to worry about feeling refreshed, she tells us. As the camera approaches a fourth woman in her bathroom, the narrator announces that a new product from the makers of Roomba called Woomba is delivering the next generation of freshness to thousands of women. Finally, that fourth woman (Tina Fey) reveals the nature of the new product saying, "It's a robot and it cleans my business, my lady business. And I like that."

Next, we see a tiny pink robot with a logo that says Woomba. It looks exactly like a shrunken version of Roomba right down to the labels on the buttons, but the mobility of this machine is astounding. Woomba can travel at high speed over any type of terrain and can even climb vertical surfaces. The sketch suggested that Woomba engaged in cleaning activities and visited locations the Roomba team had never—in our most fevered imaginings—contemplated for our robot.

I stared at my TV, mouth agape thinking, "Roomba is being parodied on national television!" Mind-blowing. In a culture where media exposure is validation, *SNL*'s two-minute sketch announced to the world, "Roomba has arrived." The mind-child we nurtured for so long and for which we had such high hopes had made good. Roomba was hereby inducted into the pantheon of American cultural icons.

Roomba, you done us proud.

NOTES

1 iRobot press release: https://media.irobot.com/2002-09-18-iRobot-Introduces-Roomba-Intelligent-FloorVac-The-First-Automatic-Floor-Cleaner-In-The-U-S.
2 *Time* magazine: https://time.com/archive/6667255/maid-to-order/.
3 Walt Mossberg, "A Vacuum Cleaner that Even a Couch Potato Could Love," 9/18/2002.
4 *Technology Review*: https://www.technologyreview.com/2002/10/09/234680/irobot-roomba/.
5 https://www.forbes.com/pictures/gllf45effm/25-the-cyber-sweeper/.
6 Early marketing and distribution arrangements were planned and negotiated by Chris Dooley and our consultant Pete Janssen, working with Winston and Jeff.
7 J-Y. Sung, L. Guo, R. E. Grinter, and H. I. Christensen, *"My Roomba Is Rambo": Intimate Home Appliances*, Proceedings of 9th International Conference on Ubiquitous Computing (UbiComp 07), Innsbrück, Austria, September 16–19, 2007.
8 They were right. Roomba did not have a powerful vacuum. The vacuum was quite literally an afterthought the focus group forced us to think. Still, our imperative was not to vacuum the floor, but to clean it. A standard vacuum wasn't necessary for that purpose.
9 I have only the highest regard for the outstanding work Per and his team did to develop Trilobite. But for a few twists of fate, the roles of Roomba and Trilobite might have been reversed.
10 Pepsi/Roomba sales: https://www.pbs.org/wgbh/nova/article/colin-angle/.

CHAPTER 19

Epilogue

WHY ROOMBA?

More than two dozen different companies and individuals made patented attempts to build a robot vacuum cleaner in the decades before Roomba, and these patents surely represent only a small minority of all attempts. Many inventors, like Frank Jenkins with his HomerR Hoover robot, built credible machines, but never applied for a patent. Given the great number of non-successes over so many years, by so many competent players— some of them very well-funded—how was it that only Roomba cracked the riddle? What special magic made that possible?

To paraphrase *Star Trek*'s Dr. McCoy, "Damn it, Jim, I'm not a business analyst, I'm just an old country roboticist." Pinpointing exactly why Roomba flourished when all others foundered is, frankly, beyond the expertise of this old country roboticist. Nor can I compose an empirical formula or plan that will enable enterprising parties to repeat Roomba's success. But I can identify critical advantages Roomba enjoyed that other pretenders to the robot vacuum crown were denied.

IMPERATIVES

We chose a good set of principles and used them to guide every decision we made.

More than anything else we focused on Roomba's price. Price drove the technical solutions we considered, the designs we developed, and every component we selected. To keep the price low, we sometimes invented low-cost alternatives to items we could have simply purchased. No component or system won a spot on the robot unless it genuinely earned its keep.

 DOI: 10.1201/9781003540489-19

Other would-be robot vacuum inventors clearly did not put price front and center. There were two reasons why we did. First, we laid the failure of so many previous robot vacuum attempts at the feet of the high prices their developers charged.[1] Despite our love for technology, no one on the Roomba team would have paid ten times the price of a manual vacuum to buy a robot vacuum cleaner had one been available. We expected that most other folks would feel the same way.

Second, we wanted to ignite the robot revolution. We couldn't accomplish that by inventing an expensive novelty for the top 1 percent. We wanted to build a robot for every floor, and to channel Henry Ford rather than Karl Benz.

We favored customer needs over roboticists' dreams. We adopted this imperative as another reaction to previous robot vacuum attempts. Many earlier inventors, we believed, had doomed their inventions by concentrating on the technology the robot contained, rather than on the utility it delivered. We Roomba developers saw ourselves not as heroic inventors of the future but as customers. We aimed to build a product that we would use in our own homes.[2]

All team members had previously experienced the various ways that complexity kills. Complexity inflates schedules, increases costs, confuses customers, and pummels reliability. Thus we considered simplicity another imperative and strove to realize it.

The team strongly believed that we had picked the right imperatives for Roomba. But choosing the correct target and scoring a bull's eye were different matters entirely. How were we able to achieve our aim?

TEAM

We had the right team. In contrast to the dissonance of the Clean project, members of the Roomba team worked in close harmony. Winston, Paul, Phil, and I were on the team by our choice. Chris was assigned to the team, but vetted by Phil. Eliot and Sara were hired for their skills and compatibility. By design—plus a bit of good fortune—we proved to be an especially compatible group. Our intense, shared desire was to build a robot vacuum cleaner that people wanted. No other agenda ever visited the Roomba office.

We trusted each other. Our common purpose and our personalities promoted mutual reliance. Objections to any facet of the design were taken seriously and examined. Crazy ideas were given an open-minded hearing. (I especially appreciated this aspect of the team psyche.)

Intra-team leadership was not a jealously guarded perk, but a role taken up and passed on as circumstance enjoined.

We were innovative. Rather than accept standard-but-incompatible solutions for power and cleaning, we reimagined the robot cleaning problem from the ground up. The many unique problems we needed to solve to birth Roomba demanded creativity and copious invention.[3]

Our management was savvy. Given the nature of our mission and team, Winston understood that a traditional, hierarchical captain-and-crew management model would have, like Clean, taken on water and sunk beneath the waves. It was unavoidable that a successful team would be composed of creative types—cantankerous cusses notoriously difficult to manage. Speaking as one of those cusses I can attest that creativity flows from internal motivations like passion, curiosity, and purpose. While corporate commands and instructions based on external motivators like raises and promotions typically fail to inspire radical innovation. Winston, abetted by Phil, worked overtime to thread the needle, to find ways to interest us in solving problems he knew Roomba (and the business) needed but that we hadn't spontaneously taken up. He always succeeded. The focus group vacuum and the late-project cleaning test were prominent examples.

SWEET SPOT

At the time we developed Roomba, iRobot occupied a sweet spot. Our location on the map was optimally situated between the neighboring communities of resources and freedom.

Typically, the smaller the company, the fewer the resources but the greater the freedom. Basement tinkerers have complete freedom but scant assets. These iconoclasts can address any problem they choose and solve it in any odd way they like (remember the Jonas brothers' unique chameleon robot Televac?). No one else is there to insist that they be reasonable, so there are no limits to their inventiveness.

Unfortunately for these folks, resources in the form of cash, space, equipment, and assistance—technical and otherwise—are typically in short supply. Carrying a significant idea all the way from drawing-on-a-napkin to product-in-a-store is exceedingly challenging. And, when that idea concerns something as complex as a robot, challenges multiply. Usually the resources required are far more than a lone inventor can muster.

At the other extreme are long-established corporate giants. Here resources are abundant, but the freedom to use them is tightly constrained. Justifications, documentation, approvals, oversight, adherence to guidelines, performance reviews, and all the other standards of good practice conspire to harry and kill immature ideas—irrespective of their merit and promise. Newly born ideas fare the worst, as they are immature by definition.

In the year 2000, despite ten years as a startup, iRobot had not yet found a compelling business model. We had no popular products to sell. Thus, freedom endured.[4] One consequence of this was that decisions could still be made quickly. It took only minutes for our CEO to decide— on his lonesome—that the company would commit to Roomba's initial development.

Our team also enjoyed rare freedom. We were allowed final authority over the design. As there were no arbitrary external constraints, we were able to follow solutions to problems wherever the evidence led. After learning something new, if need be, we could decide immediately—consulting no one—to change our design in a radical way.

We were free from the lethal beneficence of review committees and their ilk, ever poised to spew inertia onto their hapless charges. (Although we, especially Eliot, worried that such a beast could slither out at any moment. This quickened our pace.) Despite enormous resources and a mandate from the top, Electrolux took ten years to launch Trilobite. We took less than three years to deliver Roomba.

We had the resources we needed. Enough cash meant we could pay the then-exorbitant prices for 3D printing, and we could, for example, hire a consultant to design the exacting shape of the centrifugal blower we suddenly needed to enable the vacuum.

We didn't have more resources than we needed. Perversely, (as I see it) too much money is just as debilitating as too little. As I was forced to discover during childhood, limitation sparks innovation. When you can solve any problem by opening a catalog, there's no need to open your mind. But having just enough money—if you're careful!—establishes a helpfully parsimonious mindset. One that's always on the lookout for simplifications and less costly ways of doing things.

Earlier in iRobot's trajectory insufficient funds prevented us from attempting Roomba. Later, insufficient freedom would have prevented Roomba's success.[5]

LUCK

Everyone who achieves a modicum of success wants to believe that their triumph is explained solely by their own extraordinary intelligence, ability, and tenacity. I'd like to believe that too. Sadly, the evidence screams otherwise.

I was personally lucky in many ways, not least because the MIT AI Lab was hiring just at the time I needed a job. And soon after I settled in, Prof. Rod Brooks obligingly invented the behavior-based programming paradigm that made Roomba possible (or at least much simpler than it would otherwise have been). The Lab's running out of money and Denning Mobile Robotics' disdain for small robots, both painful at the time, were ultimately lucky because they landed me at iRobot.

iRobot and the Roomba team were extremely lucky in that a very long list of earlier, highly capable competitors—*who should have given us no chance to be first*—instead facilitated our dreams by graciously self-destructing. In a race between a field of venerable, fully capitalized corporate behemoths, each with industry expertise, and a deep bench of technical talent versus a pipsqueak upstart, with little cash and no relevant experience, who would you bet on to establish the consumer floor-cleaning industry? The smart money would go with the big guys, but it was the 100-to-1 long shot that won the race.

Had SC Johnson not supplied our early funding, iRobot might have decided that Roomba was still financially out of reach and halted development. And if SCJ hadn't abandoned the project when they did, Roomba might have limped into the marketplace hamstrung by an annoying and unnecessary consumable.

If Pepsi hadn't aired their fanciful (and still baffling) robot vacuum commercial when they did would Roomba have faded after our initial splash? Knowing how fickle the buying public can be, that seems a good possibility.

And, in one of the luckiest breaks of all, customers forgave Roomba its early reliability issues. The sort of problems that initially plagued Roomba would simply have doomed any normal product. The patience of our early customers still amazes me.

These are just a fraction of the instances where the simplest adverse twist of fate would have rendered Roomba a footnote in someone else's book, if that.

COMPETITION

We on the Roomba team regarded Roomba as a great idea. We figured that as soon as we showed the world how it was done, fierce competition would swiftly emerge.

But it didn't. Roomba had the market mostly to itself for a few years after launch. The first competitor I remember seeing was a knockoff produced by a factory in China. We bought one and examined it carefully. This robot had a shell different from Roomba's but everything else was the same. The plastic parts were identical—as if the knockoff company had somehow gotten access to Roomba's CAD files. Probably they did.

Software was a different story. We had taken steps to protect our code but that only went so far. We knew that a determined competitor would be able to extract the code from Roomba's microprocessor and use it in their own machine. But that is not what the knockoff had done. Running their robot, we could see immediately that they had written their own software. And it was lame. The knockoff robot didn't respond robustly to common challenges, it got stuck easily. We were confused. Had the knockoff's maker truly been unable to crack our weak protections, or (more likely) did they just see software as the easy part—not worth the effort to decode and copy?

Electrolux continued to sell Trilobite. They even updated it in 2004. But eventually, the company acknowledged the unlikely drubbing David had given Goliath, and Trilobite retired to a museum.

A few years down the road, serious competition did begin to develop. iRobot sued the makers of one copycat robot, preventing its importation.[6] Patent protection for Roomba was strong. But, as I had learned long ago, there is rarely just one way to accomplish a task or build a robot. As companies found workarounds, more and more vacuum cleaning robots began to enter the market. At last count, about 25 different companies were making robot vacuum cleaners for consumers.

Over time iRobot introduced lots of new models. But it was with great satisfaction that I regarded iRobot's new Roomba variations and the emerging competitors. Everyone more or less copied our team's original design. All the robots had cliff sensors (some were implemented using Sharp sensors so as to sidestep the patent), they all had mechanical bumpers, they all had at least one side brush, they were all about Roomba's size, they all relied mostly on a carpet-sweeper mechanism to accomplish cleaning, and at least initially they were all round. Sincere flattery indeed.

ELECTROLUXIFICATION

Soon after our robot became successful, Roomba team members started to leave iRobot.

Every startup begins with a vision—a bold new approach that founders believe will bring customers flocking to their door. But that vision is necessarily unproven, so a startup must swallow a large dose of existential risk in its pursuit. The rate of infant mortality is horrendous. Within just a few years of their beginning most startups end—the majority in failure; a lucky few move on to become established businesses.

iRobot's founding vision was that a behavior-based approach to robotics would trigger the robot revolution. Why wouldn't success in the academic realm transfer to the commercial? (We naïvely thought that question was rhetorical.) As first movers, we would lead a transformation, developing new robotic applications one after another. And, unlike the dated vision from the 1950s, our robots would ease the burdens not just of America's housewives but of everyone! Along the way, we'd transform a bunch of industries.

The journey took far longer than any of us expected. Rather than either succeeding or failing within a few years, iRobot remained a startup for a dozen. During this time, risk was constant. Despite a gradually improving financial situation, for a decade-plus the company lacked a dependable income stream. In its place was an unpredictable series of contracts and grants. Winning the latest was no guarantee of winning the next.

Then came Roomba. And at about the same time PackBot.

Fortunes changed. Finally, we had products to sell that people, in increasing numbers, wanted to buy. What a concept! This gave the company a crank to turn. Spend X on promotion, expect Y back in revenue. For the first time iRobot's income became relatively stable and predictable.

Those of us who'd joined in the early days shared our CEO's oft-stated aspiration that iRobot would become *the* robot company. Any and all new mobile robot applications would be fair game. Our destiny was to invent the lion's share of this flood of robotic products. That history primed me to believe that the company would be eager to repeat Roomba's success, and that the reliable income stream provided by Roomba and PackBot would translate to many more shots on goal. An abundance of both freedom and resources would let us probe new possibilities and new applications. Developing speculative new robots would be our calling card.

But I was wrong. I hadn't understood how startups become established corporations any better than I initially understood how carpet sweepers pick up dirt. Success made the Electroluxification of iRobot all but inevitable. Aspects of that process had begun even before Roomba.

As innumerable business books explain, real innovation is next to impossible for established companies. Startups and their adult siblings have missions that are different and incompatible. And it's not just their missions that differ. The outlook of employees most comfortable at each establishment are also dissimilar; they diverge in their motivations and tolerance for uncertainty and failure.

A startup has but one imperative: find a business model that works. That is, the task is essentially one of exploration. The situation matches that of the wizened old prospector, donkey in tow, scouring desolate lands for fortune. This lonely seeker follows no chart—there is no map to gold that has not yet been found. Neither rule nor process nor policy handbook can reveal the whereabouts of hoped-for treasure. The prospector is free to, and indeed must, improvise.

But if the prospector beats the odds and strikes gold, everything changes. Those who grubstaked the prospector prescribe that exploration now transform into exploitation. This second act, the established business, banishes improvisation; it follows only a well-researched script found in textbooks, case studies, and business school classes. Executives, managers, and accountants join the cast in leading roles, and the prospector, who has trouble reciting lines written by others, is reduced to a cameo. (The new stars wager that the vein they mine will never play out.)

Established businesses insist on rules, processes, adherence to policy, and other things (things we of the prospector-persuasion find boring and stifling) not because of myopia or moral failing but because these practices genuinely work. Conventional operating imperatives enhance profits. They please shareholders. They maximize market success by ensuring that the company develops only well-researched products that existing customers are known to want. No speculative products need apply—likewise the cantankerous contrarians who might build them.

Still, this near-inevitable scourge of success is uncomfortable to those of us who delight in discovery rather than corporate dictate. As Roomba sales continued to climb, I set off in search of other Roombas—and was mystified by the company's disinterest in my quest. Four years on, reality became undeniable: I could pursue other revolutionary products only at another startup.

A few at iRobot maintained that Roomba's success was a fluke. They pointed out, correctly, that our team was composed of oddballs and

202 ■ Dancing with Roomba

misfits who were managed in a suspiciously unconventional way, and that our product wasn't even a new idea. Surely, they argued, a positive outcome was eked out in spite of the Roomba team, not because of it. It followed that a team captained by a traditional manager and crewed by traditionally motivated employees would have produced an even better outcome—and only that conventional structure could ensure success going forward.

The depth of such feelings was telegraphed during a meeting of the board of directors sometime before launch. A Roomba prototype was demonstrated to the board; a manager from one of the other projects—a proponent of the conventional view—was present at the meeting. During the demo he harried the robot as it ran, seeking to make it fail. He dragged the heel of his shoe across the tile floor, leaving a black mark, and pointed out that Roomba was unable to erase it.

The marketplace's enthusiastic response to Roomba was a source of great pride and satisfaction to our team. We felt that this evinced proof of our historic feat. It came as a disheartening shock that the company seemed not to share this view. iRobot acknowledged our work by distributing cash bonuses equivalent to the value of five robots to team members, and we were granted a discount toward the purchase of the robot we had designed. Then our team was disbanded.

Winston had been pushed out shortly before the project was done. Eliot and Phil left the company the year after launch, Sara before that. Paul and I hung on for four years, departing to found another startup. Chris exited a few years later.

But the lily needs no gilding. Roomba's singular accomplishment eclipsed any recognition-related disappointment we felt as members of this remarkable team.

CODA

How do you build a practical robot vacuum? That simple riddle intrigued, perplexed, and stymied scores if not hundreds of inventors for nearly 50 years. Inspired by science-fiction writers and futurists, convinced that the answer was within reach, many actively sought a solution. Over the years, numerous products were announced and demonstrated. But most faded before they were launched, and none found purchase in the marketplace.

Then, Roomba cracked the enigma. Our robot delivered on the decades-old promise and birthed a new industry. Roomba's design became the archetype and benchmark for all that followed. It launched a million memes.

Working on Roomba, bringing it into being, and watching it flourish has been a sublime honor. Rarely, over the entire course of a professional career, is one privileged to participate in a phenomenon like Roomba—exhilarating, humbling, confusing, and joyous sometimes all at once. It has been the adventure of a lifetime, a most unexpected escapade for a kid from the country.

For anyone looking to court an experience like Roomba, the best advice I can offer is to follow your interests, be alert to opportunities, and endure the setbacks. I hope your journey will be, is, or was as fun and fulfilling as mine has been.

NOTES

1 Most developers never brought their inventions to market and so never had to divulge a firm price. However, we could often guess the lowest price they'd be able to charge given the technology the device contained.
2 Since we started prototyping the robot I've always had a Roomba in my home. I run it regularly—but of course, not as regularly as I should.
3 iRobot has been granted on the order of 100 Roomba-related patents. I'm the inventor or co-inventor with other team members on about 80 of these.
4 As Janis Joplin reminds us, "Freedom's just another word for nothing left to lose."
5 For at least two decades after Roomba and PackBot launched, iRobot gladdened the hearts of shareholders by mostly prospering, growing, and demonstrating fiscal prudence. But the company delivered no further revolutionary products.
6 In 2005 Koolatron Corporation settled the lawsuit, agreeing not to bring their KV-1 floor-cleaning robot into the United States.

Appendix: AI and Robots

BEHAVIOR-BASED PROGRAMMING WAS THE final technological advancement needed to make consumer floor cleaning robots practical. My colleagues and I at iRobot believed that behavior-based programming was *the* method that would open the floodgates of robot applications. Today, more new types of robots are trickling down the spillway than in the years before Roomba. But our grander hopes remain unfulfilled.

Robotics is artificial intelligence embodied. And in the realm of AI, my unbridled optimism put me in good company. Since its inception as a discipline[1] the difficulty of the sorts of problems AI took on and the time needed to find solutions have been grossly underestimated at every turn.[2]

Over the years many different techniques, each billed as the "solution" to AI, have taken their place in the limelight. Each in turn received the adulation of the crowd but was later demoted to a supporting role without ever delivering the promised grand finale. A few former divas include logical theorem proving, expert systems, statistical inference, and fuzzy logic.[3]

Currently performing at center stage are artificial neural networks (ANNs). Nurtured by a devoted few over many lean years, recent breakthroughs have transformed this once often-disparaged approach into the most impactful and consequential technique in AI's seven-decade history. ANNs underpin deep-learning, generative AI, and the many applications currently exciting hope and horror among the populace.

Notwithstanding its role in self-driving cars, AI is mostly confined to the innards of computers. But is this about to change? Will modern AI finally enable Asimov's dream of the general-purpose robot—a dexterous machine able to match or exceed human performance at any task? The question is a popular one, asked in robot labs around the world.

AI

To achieve artificial intelligence one early idea[4] was to imitate living intelligence. Brains enable thinking; so, we should build a mechanism that replicates the key aspects of living brains.

Neurons wired to form a network are the foundation of organic brains. Most neurons involved in thinking have a central body, a single axon for output, and many dendrites for input. Synapses connect a dendrite from one cell to the axon of another. The cell body sums the signals from all the cell's dendrites. When that sum exceeds some threshold, the neuron fires, sending a pulse down its axon. The strengths of synaptic connections affect the sum and thus neuron firing. Learning is accomplished by adjusting connection strengths.[5]

Artificial neural networks mimic this. ANNs are commonly depicted as a matrix of circles—the cell bodies. Each column of circles is thought of as a layer. The leftmost column represents the input layer, the rightmost the output layer. Some number of "hidden" layers intervene. Lines, representing dendrites, connect the neurons. The output (the axon is not explicitly represented) of each neuron on one layer connects to inputs of neurons on the next layer. Computation proceeds as each neuron multiplies the value of each of its inputs by stored weights and then accumulates those products into a sum. It then computes a non-linear output value based on that sum.

Suppose we wish our ANN to identify which of several possible items is present in an image. To do this we can connect each input neuron to one pixel of the image. We then assign one output to each possible item. Say we associate the first output with "cat," the second with "aardvark," and so on. After running our neural net calculation, the output with the largest value tells us which item was in the image. Normalizing[6] the outputs let us think of the values as probabilities.

Training the network enables identification. We begin with a data set of many images, each labeled with the item present in that image. Suppose we present an image containing a cat to our ANN. In this case, the output neuron representing "cat" should have a value of one (100 percent); all other outputs should be zero. When this is not the case, we adjust all the weights in the network (there can be millions) in a way that increases the "cat" output and decreases the others. We repeat the process for all the labeled images in our data set. With enough training, the network learns to recognize cats and other items.

ANN researchers wandered in the wilderness for many years in part because it was so tough to figure out a workable way to adjust those potentially millions of weights. (Many of us were convinced they never would!) But a method called *back-propagation*[7] eventually led the faithful[8] to the promised land.

The seductive aspect of ANNs is that all they need is data. Previously, to identify a cat in an image researchers had to write complex and challenging programs. Exhaustingly they would apply multiple advanced mathematical techniques, heuristics, and a little bit of learning to an image to eventually arrive at a sometimes-correct diagnosis of "cat."

But with ANNs, in a sense, the computer programs itself! All the "programmer" needs to do is show the ANN thousands or millions of labeled images. It figures out the rest. The internet came along at just the right time to supply the mountains of needed data for free.

In decades past there were competitions between researchers to determine whose program could most accurately identify items in images.[9] But around 2012 the contest started being won exclusively by ANNs.[10] Humans with their clever, laboriously constructed programs were left in the dust.

That stunning success aroused enormous interest. Researchers flocked and money flowed to ANNs. Different ways of using the technique were developed. It turned out that not only could ANNs identify images, but they could generate them too. And they could translate documents, interpret speech, and write college essays.

ROBOTS

The latest AI techniques appear to have largely solved previously intractable computer vision and natural language processing problems that researchers had wrestled with for half a century. Perhaps the same techniques are about to "solve" robotics. Is it now the case that a robot can learn to perform any task if we just feed it enough data? Or will additional breakthroughs be needed? Researchers have voiced opinions on both sides of that divide.

DATA IS ENOUGH

One of the most often cited reasons to believe that we've already crossed the Rubicon is the stunning capabilities ANN-based methods have demonstrated to date. If an AI can read handwritten text, interpret and generate natural-sounding speech, and create photorealistic images of fanciful things it's never seen, then why wouldn't it be able to direct a robot with similar aplomb?

Another supporting indicator is the nature of the algorithm. Artificial neural networks rest on a simple foundation. Neurons just multiply numbers from other neurons by a weight, add them together, and output a number based only on that sum. Systems can be made as large or as small as needed. They can be trained on as much or as little data[11] as you fancy. That is, ANNs scale well. Historically, simple algorithms that scale well have performed better than clever but complex algorithms that don't.[12] Furthermore, more data and more computer power are always on the way. Bigger networks trained on more data (presumably) equals better performance.

The ability to translate a spoken language versus identify an item in a static image appear quite different. But the fact that both can be competently addressed by using the same technique suggests similarities at some basic level. The ability to direct a robot to accomplish different complex tasks, say mechanical watch repair and rodeo bull riding, also appear different. But surprisingly, it may be that at a deep level all practical tasks are similar and can be solved by the same technique.

Finally, many real-world problems would be so much simpler to solve robotically, if only robots had a modicum of common sense! An AI exhibiting common sense has yet to be convincingly demonstrated.[13] However, it seems plausible that common sense may emerge from larger models compiled with ever more data.

IS IT THOUGH?

ANN-based AI systems have an annoying habit of hallucinating[14]—they sometimes return results that are nonsensical or wrong. It's proven hard to eliminate this trait,[15] likely because current AIs are based not on reasoning but statistics.[16] Modern AIs are in the business of computing the most probable answers based on their training. But sometimes the most probable answer is not true. Then they hallucinate.

In worlds of text and images, the occasional hallucination might be regarded as humorous. But robots operate in the physical world and physics does not forgive hallucinations.

A frequently heard argument takes the form: If X can do A, then it must be able to do B. Some data-is-enough proponents make that case. But whether it's true depends on properties of X, A, and B that we may not know. The prowess AI has demonstrated in some fields may carry over to robotics—or not.

One practical issue remains even if a data-trained robot could perform any task: getting the data. Training an AI to identify items in images may require millions of images. These are available for free[17] on the internet. Training an AI to translate or write text required millions of lines of text, also available for free. But training a robot to perform physical actions may require similar quantities of examples of robots executing physical actions. Furthermore, the required data may be specific to the sort of robot being trained. Data generated by a two-wheeled, indoor, cleaning robot may be useless to a two-armed, four-legged walking robot intended to maintain hiking trails. Appropriate robot data mostly doesn't exist, and generating it isn't free.[18] So it may be that this route to a general-purpose robot isn't affordable.[19]

WILL SONNY[20] SELL?

In the years before Roomba, many different individuals and groups demonstrated working floor-cleaning robots. But all those enticing presentations skirted two awkward issues: performance and cost.

First, demonstrating that a machine "works" isn't enough. The appropriate standard for a product is not that it often works but that it almost never fails. A robot vacuum that cleans superbly but requires rescue nearly every time it runs will attract few customers. And a robot of any purpose that damages its user's property—even occasionally—is more likely to inspire a lawsuit than a shrug. Neither "working" machine is viable.

The stakes for a device that directly affects the physical world are higher than for a program that delivers its output to a human. A person can benefit even from an imperfect program by filtering the results—accepting helpful answers and rejecting pernicious ones. But an AI that occasionally hallucinates while controlling a robot gets no saving throw. Physical injury to persons or possessions may result.

Second, cost is determinative. A robot, controlled by AI magic, that performs a task as competently as a human has addressed only half of the problem—the easier half. To provide positive value, the lifetime cost of that robot must also be less than other solutions.[21] Otherwise, our brainy automaton will find itself standing in the unemployment line.

ANN-enabled AI is a technique proven to have enormous power. Certain existing robots already employ that method to solve parts of specific problems.[22] There will be more such robots. But whether a generic platform robot, trained only by data, will be able to perform any desired task is an unanswered question. The outcome depends on the synergistic confluence of technology and economics, not on either alone. Even if we could build a real-world Sonny, the robot's coolness and capabilities would not guarantee sales. Only if real Sonny costs less to buy and use than the bespoke products or human worker that might otherwise perform its duties will anyone take Sonny home from the robot store.

NOTES

1 The generally accepted birth of AI was the Dartmouth Workshop held in 1956.
2 In the 1960s some researchers believed machine intelligence would equal human intelligence within a generation: https://www.sydney.edu.au/news-opinion/news/2024/09/02/ai-was-born-at-a-us-summer-camp-68-years-ago.html. Self-driving cars offer another indication of AI difficulty. One of the first attempts began at CMU's Navlab in the 1980s, yet even today the problem isn't fully solved.
3 You'll find a wealth of information about AI in: S. Russell and P. Norvig (2021), *Artificial Intelligence: A Modern Approach*, 4th edition, Pearson.
4 Warren S. McCulloch and Walter Pitts (1943), "A logical calculus of the ideas immanent in nervous activity," *Bulletin of Mathematical Biophysics*, 5 (4): 115–133. This important paper showed how neurons (or at least simple models of neurons) could perform logical operations.
5 This is a simplified overview. Real neurons often exhibit more subtle trickery.
6 To normalize a bunch of (positive) numbers add them all together, then replace each number by its original value divided by that sum. The new set of numbers all have values between zero and one and if we add these numbers together, the total is one. A complete set of probabilities also have those two properties.
7 The key paper was: D. Rumelhart, G. Hinton, & R. Williams (1986), "Learning representations by back-propagating errors," *Nature*, 323: 533–536. https://doi.org/10.1038/323533a0. But after publication another 20 years, additional advances, and better hardware were needed before ANNs could achieve breakthrough performance.
8 Chief prophet was Geoffrey Hinton who (with others) received the 2018 Turing Award and the 2024 Nobel Prize in Physics.
9 https://en.wikipedia.org/wiki/ImageNet.
10 A program called Alexnet, devised by Hinton, et al. led the way; see: https://computerhistory.org/blog/chm-releases-alexnet-source-code/ and https://viso.ai/deep-learning/alexnet/.
11 Of course, depending on the problem, performance may suffer if you rely on too little data.
12 N. Kumar, "Will Scaling Solve Robotics?" IEEE Spectrum, 2024: https://spectrum.ieee.org/solve-robotics; see also: http://www.incompleteideas.net/IncIdeas/BitterLesson.html.
13 Some researchers argue that AIs *have* achieved common sense. But my inner curmudgeon is unconvinced that any real-world robot has done so.

14 Many researchers recoil from using the word "hallucinate" to describe this sort of AI error. AI systems do not hallucinate in the sense that humans do. Rather they sometimes produce inaccurate answers and generate false information. I agree that the word may evoke unwarranted expectations, but use the term here only because it has become so embedded in the popular vernacular.

15 Will they get there? Progress has been disconcertingly rapid, any statement made today about what an AI can't do may be outdated tomorrow.

16 Symbolic AI was one of the great hopes in the early history of the field. Such systems *did* base their outputs on reasoning and did not hallucinate. Unfortunately, early systems were brittle and didn't scale well. However, researchers are working to combine the best features of these systems and ANNs.

17 The data is free when no human must be paid to label it. Some online images have associated labels, and some learning techniques don't require labels.

18 Some such data is available. See: https://droid-dataset.github.io.

19 On the other hand, physical simulations just keep getting better and better. Simulations wouldn't have helped Roomba, but maybe the modern version will enable inexpensive robot training.

20 A prominent role in the 2004 movie *I, Robot* was played by Sonny, a general-purpose, humanoid robot.

21 I've built robots other than Roomba. Painful experience affirms that regardless of how well a robot performs its task, unless customers perceive an economic value superior to competing products or services, they will not adopt the robot.

22 For example, Brightpick makes a line of warehouse robots that use AI to recognize items to be picked.

Acknowledgments

I'M A TECHNOLOGIST AND—WRITING what I know—I've told Roomba's story mostly from a technology perspective. But technology wasn't close to enough. A successful product rests on a broad foundation of which innovative technology is only one essential pillar. Others included astute business decisions, shrewd financing, inspired marketing, logistical legerdemain, efficient manufacturing, and much more. My first acknowledgment and sincere gratitude goes to the champions of those other disciplines and to the many people, most of them unnamed herein, who together made Roomba possible. It took a village.

A big perk of working on the book was that it let me get back together with my Roomba teammates: Winston Tao, Paul Sandin, Eliot Mack, Phil Mass, Chris Casey, and Sara (Farragher) Massari. They supported my battered natural neural net by reminding me of events I'd forgotten or misremembered, they supplied stories or angles I'd missed, and they provided manuscript reviews.

Helen Greiner, iRobot co-founder and serial entrepreneur, helped out through discussions about the old days and by reviewing the manuscript. My favorite faux bon vivant, Jeff Ostaszewski, contributed insights and a review.

My old boss, Prof. Tomás Lozano-Pérez (who unknowingly assisted Roomba's birth by indulging my mobile robot fascination when I should have been working more on his projects), helped out. He provided pointers and information about contemporary artificial intelligence and reviewed the text. I received additional help from Prof. Holly Yanko and my early iRobot colleague, Chuck Rosenberg.

Numerous friends made the book better by graciously contributing their time to review the text and suggest changes. They include Dr. Rick Shafer, Prof. Lesley Sharp, Anne Jones, Clara Vu, and Dr. Gary Galica.

I am deeply indebted to Per Ljunggren, Trilobite project leader, who shared many fascinating details about Trilobite's development with me. Electrolux engineer Ulrik Danestad also helped in that way.

Thanks go to the crew at CRC Press, Lucy McClune, Matthew Shobbrook, and Randi Slack, for their part in making *Dancing with Roomba* real. Working with them has been a pleasure.

Finally, heartfelt thanks goes to my family for their love and support: my wife Sue Stewart, daughters Kate and Emmeline, and my sister Karen McDonald-Bowman. They all went beyond the call of familial ties helping me to improve the manuscript by reviewing it at various stages and offering astute suggestions.

Index

Page numbers followed by "n" refer to notes.

For Product Safety Concerns and Information please contact our EU
representative GPSR@taylorandfrancis.com
Taylor & Francis Verlag GmbH, Kaufingerstraße 24, 80331 München, Germany

www.ingramcontent.com/pod-product-compliance
Lightning Source LLC
Chambersburg PA
CBHW061159220326
41599CB00025B/4541